Norbert Wiener

A Mathematician among Engineers

Norbert Wiener

A Mathematician among Engineers

José María Almira

University of Murcia, Spain

World Scientific

NEW JERSEY · LONDON · SINGAPORE · BEIJING · SHANGHAI · HONG KONG · TAIPEI · CHENNAI · TOKYO

Published by

World Scientific Publishing Co. Pte. Ltd.

5 Toh Tuck Link, Singapore 596224

USA office: 27 Warren Street, Suite 401-402, Hackensack, NJ 07601

UK office: 57 Shelton Street, Covent Garden, London WC2H 9HE

Library of Congress Cataloging-in-Publication Data
Names: Almira, José María, author.
Title: Norbert Wiener : a mathematician among engineers /
    José María Almira, University of Murcia, Spain.
Other titles: Norbert Wiener. English
Description: New Jersey : World Scientific, [2023] | "The first edition of the present book
    originally appeared in Spanish as Norbert Wiener. Un matemático entre ingenieros in
    2009"-- foreword. | Includes bibliographical references.
Identifiers: LCCN 2022030928 | ISBN 9789811259364 (hardcover) |
    ISBN 9789811259371 (ebook) | ISBN 9789811259388 (ebook other)
Subjects: LCSH: Wiener, Norbert, 1894-1964. | Mathematicians--United States--Biography.
Classification: LCC QA29.W497 A7613 2023 | DDC 510.92 [B]--dc23/eng20220919
LC record available at https://lccn.loc.gov/2022030928

British Library Cataloguing-in-Publication Data
A catalogue record for this book is available from the British Library.

For any available supplementary material, please visit
https://www.worldscientific.com/worldscibooks/10.1142/12919#t=suppl

Desk Editors: Nimal Koliyat/Rok Ting Tan

Typeset by Stallion Press
Email: enquiries@stallionpress.com

Printed in Singapore

To Prof. John B. Anderson, with admiration and dearness

Foreword

Norbert Wiener (1894–1964), along with G. D. Birkhoff, was probably one of the first American mathematicians to achieve international prestige. He was a child prodigy who graduated when he was only fourteen years old and, at eighteen, defended his doctoral thesis in mathematical logic at Harvard University and obtained a post-doctoral fellowship to do research with the famous English logician Bertrand Russell in Cambridge, England. Indeed, Wiener's Ph.D. thesis was related to Russell and Whitehead's *Principia Mathematica*. After several disagreements between the two men and under the influence of the also famous English mathematician G. H. Hardy, Wiener switched from philosophy to mathematical analysis and got a teaching position at the Massachusetts Institute of Technology (MIT) in 1919. This is how he began a brilliant career as a mathematician, the starting point of which was the precise mathematical formulation of Brownian motion, a subject on which he worked between 1920 and 1923. With this research, Wiener became one of the initiators of the theory of stochastic processes and a precursor of the theory of probability in infinite-dimensional spaces, and with it he managed to convince Europe about his worth as a mathematician, but not the United States. The next steps taken by Wiener, until he achieved the recognition of his American colleagues (which would come to him in 1933 with the award of the Böcher Prize), were the solution of the Zaremba problem, a rigorous foundation for Heaviside's operational calculus, the creation of generalized harmonic analysis, and the development of a profound tool motivated by the Tauberian theorems, a theory from which he would obtain a new proof of the most important result in analytical number theory, the prime number theorem. Not content with these achievements, Wiener would address many other issues, including ergodic theory, wave filter theory, information theory (giving his own definition of entropy), and prediction theory. Taking all these ingredients together, he ventured to create a new scientific paradigm: cybernetics.

Although it reached the highest levels of abstraction, his mathematical work was always motivated by physics, engineering, or biology. For example, generalized harmonic analysis is a tool of enormous significance in studying weak currents, a part of electrical engineering the importance of which has been increasing throughout the 20th century and continues today. Of particular importance are the problems of filtering (that is, the separation of noise from a message) and prediction theory, and, in both cases, his contributions were vital.

In addition to being a scientist of the first order, Wiener was an intellectual and a rebel. When the United States dropped its first atomic bomb on Hiroshima, he made a public call to all American scientists to abandon any research with possible warlike applications. He himself systematically refused to participate in projects with military funding, which isolated him from his colleagues. This, on the other hand, allowed him to spend time writing several popular books that made him a public figure.

My objective in this book is to explain the life but also the mathematical work of Wiener. I would like to think that this biography will be of interest to mathematicians, physicists, engineers, and even biologists, physicians, and philosophers. Wiener's work is so interdisciplinary and affects so many spheres of thought that such a claim is by no means far-fetched.

The first edition of the present book originally appeared in Spanish as *Norbert Wiener. Un matemático entre ingenieros*[a] in 2009. The present edition can be considered a revised expansion of the original

---

[a]J. M. Almira, *Norbert Wiener. Un matemático entre ingenieros*, Ed. Nivola, 2009.

one, since several major modifications have been added, including the last section of Chapter 5, devoted to the militarization of science in the USA, and an extensive appendix where Wiener's and Shannon's contributions to the rise of a Digital World are explained and confronted. Thus, this book can be considered in many respects a new book.

About the Author

 **José María Almira** is a Professor (Associate) of Applied Mathematics at the University of Murcia. His main research interests are Functional Equations, Functional Analysis, and Approximation Theory. He has published more than 75 research articles in prestigious national and international journals and 14 books. Eight of his books are scientific biographies, five of which have been translated into Italian and French and published in six countries. In addition to teaching and research, he has dedicated a good part of his professional career to scientific dissemination. In particular, during the last two years, he has been the Scientific Director of the collection of books "The mathematics that transforms our world," the first edition of which has already been published in Italy (in Italian) and contains 35 titles.

Contents

Appendix: Wiener, Shannon, and the Rise of a Digital World — 233

Chapter 1

Childhood and Youth: From Child Prodigy to Doctor in Philosophy

A Father and Mentor

It is impossible to understand who Norbert Wiener is without knowing some details of the life of his father, Leo Wiener, and the influence and admiration that Wiener showed for him, so much so that, in the first volume of his autobiography, which appeared in 1953 under the title *Ex-Prodigy: My Childhood and Youth*, Wiener dedicates the first two chapters almost entirely to talking exclusively about his origins in general and his father in particular.

Bertha Kahn (1867–1964) and Leo Wiener (1862–1939), parents of Norbert Wiener. Courtesy: MIT Museum.

Leo Wiener (1862–1939) was born in Byelostok (Białystok, in Polish, is the current name of the city), then a Russian city, and today a Polish one, the capital of the Podlakia Voivodeship. He was the son of German Jewish immigrants. According to the *Encyclopedia Britannica*, 1911 edition, two-thirds of this city was inhabited by Jews. Located in an area politically and socially influenced by Russians, Poles, Belarusians, and Latvians, Byelostok was an intermediate

1

point on the railway route between Moscow and Warsaw, the latter being one of the cities where Leo Wiener was educated (along with Minsk and Berlin). Thus, not only did he receive the aforementioned cultural influences, but also he learned all the languages of the surrounding regions by being educated in them.

Leo Wiener was a descendant of Aquiba Eger (Chief Rabbi of Posen from 1815 to 1837). According to his family tradition, he was also a descendant of Moses Maimonides, the famous Cordovan rabbi and philosopher who lived during the 12th century.

This cultural hodgepodge fueled his interest in languages and forged an astonishing cover letter for this 18-year-old Russian who arrived in New Orleans in 1880 "with 50 cents in his pocket." He spoke perfect German (the language of his family, in which he was educated in a Lutheran school) and Russian (the language of his native country). Moreover, he had learned French and Italian as educated European languages of the time and Polish during his stay in Warsaw (where, after meeting Zamenhof, the inventor of Esperanto, he was one of the first interested in and scholars of this artificial language). Also, during his stay at the Berlin Polytechnic, he learned modern Greek and Serbian. Furthermore, Wiener defined his father as an excellent student of Latin and classical Greek. He also went on to say of him, about his command of English, that "my impression of his English is that it showed its foreignness more in an excessive precision of diction and of vocabulary than in any other way." Not surprising, since he had learned it during the two-week boat trip to New Orleans by reading Shakespeare's works as well as studying Spanish during that same period.

During his Berlin years, Leo Wiener came into contact with student circles of a leftist nature. Curiously, Wiener, in his 1953 autobiography, describes these groups as "humanitarian" probably due to political tensions existing in 1953, mainly due to McCarthy and his particular "witch hunt" against everything that "smelled of communism."

Thus, Wiener comments that his father, after attending a series of meetings and conferences, with a reinforced vein of Tolstoianism, "decided to deny drink, tobacco and meat for the rest of his life." This vegetarian-socialist environment led him to make the decision of his life: to emigrate to the United States to found a vegetarian commune there.

Despite being a man of letters, Leo was very interested in mathematics during his education and, although he did not make use of that knowledge in any of his works, he had a deep interest in this field throughout his life. Norbert claims that "during school he received much of his mathematical training from his father" and also that "it would have been better for him [Leo Wiener] to choose the field of mathematics rather than philology."

Already in New Orleans and after an unsuccessful period where various jobs in factories and farms followed one another, Leo Wiener moved to Kansas in search of more intellectual activity. It was there that, following a sign posted in a church, he joined the local Irish society that taught courses in Gaelic. In time, he earned the affectionate nickname of "The Irish Russian" in that city because he became a teacher and the director of the society. In addition, he worked as a language teacher in a Kansas high school, later moving to Columbia, where the University of Missouri hired him as an assistant professor of modern languages and German. In 1893, he married Bertha Kahn, the daughter of Henry Kahn, a Jewish warehouse owner in St. Joseph, Missouri, and after one year, Norbert Wiener, the first of six siblings, was born.

Leo Wiener was a prolific writer and eminent philologist and is not without some fame within the area today. Thus, among his most famous works is an anthology of Russian literature, the books *An Interpretation of the Russian People*, *Commentary on Germanic Laws and Medieval Documents*, and the edition and translation (in two years!) of *The Complete Works of Tolstoy* in total 24 volumes. After passing through the University of Missouri, he became Harvard's first Jewish associate professor in 1896 (and later, in 1911, full professor), teaching Romance and Slavic languages, finally retiring as professor emeritus in 1930.

Birth and Early Years

Our protagonist was born in Columbia, Missouri (United States) on November 26, 1894. As he later defined himself, Wiener was "seven-eighths of a Jew" and, although he would not discover his true origins until he was 15 years old, he was nevertheless educated in this culture, which marked him. His name, Norbert, was chosen by his parents for being one of the two characters in the play

On the Balcony by the English playwright and poet R. Browning (1812–1889).

In his autobiography, Wiener does not spend as much time on his mother as on his father. While with his father, he is incredibly descriptive (and, in a way, flattering), also highlighting all the "intellectual pressure" suffered during his education, his mother Bertha remains a very secondary character in his life, assuming a rather standard role in Wiener's upbringing. He tells us that she was "a small woman, healthy, vigorous and vivacious."

Although he was born in Columbia, in the heart of the United States, he spent his early years in New England, as his father had obtained a teaching position in Cambridge (Massachusetts). This would be the state where Wiener would grow up and to which he would curiously return years later when he obtained a teaching position at the Massachusetts Institute of Technology (MIT).

Norbert Wiener when he was two years old. Courtesy: MIT Museum.

Except for a brief period in 1901, Wiener was educated entirely by his father before entering secondary school. Leo Wiener had decided that his firstborn was going to be a genius. Not a genius by chance, but a deliberately fabricated one. Thus, Norbert Wiener, who was born with great expectations, was already able to read perfectly at the age of three and very soon had to study Greek, Latin, mathematics, physics, chemistry, etc. Not surprisingly, he was growing up in a purely intellectual environment:

> I was brought up in a house of learning. My father was the author of several books, and ever since I can remember, the sound of the typewriter and the smell of the paste pot have been familiar to me. (...) I had full liberty to roam in what was the very catholic and miscellaneous library of my father. At one period or other the scientific interests of my father had covered most of the imaginable subjects (...) I was an omnivorous reader.

In 1898, when Wiener was four years old, his sister was born, named Constance. In this case, they chose the name of the other main character in Browning's play. A brother of Wiener was born in 1900 but died shortly after.

Recalling his childhood, Wiener is quite impressionistic with details that, although from a biographical point of view do not provide much information, they do help to perfectly frame the succession of events that led to shaping his life. One of these carefully listed details is the different readings that marked him. Until 1901 — when he was still a seven-year-old boy! — he read all kinds of books on a wide variety of subjects. Afterward, he began to form his scientific character. Among his early books is a copy of J. G. Woods' *Natural History*, which a friend of his father gave Norbert when he was just three. At the same time, he got into his hands a volume of elementary science that he tried to understand despite his young age. Finally, he discovered, at the age of six, the *Advanced Arithmetic* by G. A. Wentworth (1835−1906) (the mathematics textbook used in high school at the time). In his own words: "I read ahead into the discussion fractions and decimals without any great difficulty."

First Trip to Europe

In 1901, the family made a long trip to Europe, first passing through New York. So, during the spring, they visited the father's brothers, who lived in New York. In particular, Wiener remembers his uncle Jake, whom he describes as a strong man "even wider in the shoulders than my father" (whom he had referred to as someone possessing this characteristic). This recollection coincided with reality because "Uncle Jake" was a cultured sportsman who had risen to number three in the United States in gymnastics, specializing in parallel bars.

Wiener also describes his two aunts Charlotte and Augusta as "exceptionally intelligent" women, with abilities and careers comparable to his father's, who had maintained much of the Russian culture as they become businesswomen, along with some French culture due to a stay in Paris for a few years.

He also remembers the gifts they gave him at that time, surely seeking the entertainment of a child about to embark on a very long trip to Europe (something that he probably still could not understand). He received three educational games: "Have fun with electricity," "Have fun with magnetism," and "Have fun with soap bubbles." The first two, presumably intended for older children (especially if we consider that this happened in 1901, and remember that he was only seven years old). On many occasions, it seems that exchanging a ball for more or less didactic books or gifts was forming in Wiener a scientific spirit, fundamentally marked by curiosity and searching for the why of things.

The trip to New York was only the first stop on the European journey that would take the family from "Cambridge to Cambridge, via New York and Vienna" (as stated in the fourth chapter of Wiener's autobiography). Days later, the family took a Holland America Line ferry bound for Rotterdam. After a very long journey, in which Wiener had the opportunity to practice his German and learn a few words in Dutch thanks to the announcements to the passengers, they finally reached their destination. From there, they first set out by train to Cologne to visit a cousin of his father.

However, the trip to Europe was not aimed at a family visit. It was more a series of interviews that Leo Wiener wanted to have with some relevant characters for his work as a language professor at Harvard. Thanks to him, and his native knowledge of German, they were able to move around different countries in such a way that Wiener went so far as to say that "to visit Europe with my father was to see it through the eyes of the European." Thus, they soon set course for what would be their next destination: Vienna.

The visit to Vienna had a fairly clear objective: to see Karl Kraus (1874–1936). Kraus was an Austrian journalist and writer known for his essays and poetry. He was also relevant for the satirical view he had of German culture, and Austrian and German politics. Wiener does not know the exact subject of the meeting Kraus had with his father but speculates that it may have had to do with the literal translation into German of the Yiddish poems of Moritz Rosenfeld, the Jewish poet "of the workers' class."

Due to a plague of insects that ravaged Vienna and populated the beds, biting everyone who got in front of them, the family moved to the small town of Kaltenleutgeben as guests in a shoemaker's house. Wiener remembers being able to communicate with the other children in the house in a language that was neither English nor

German. The deep family atmosphere of study in the many languages in which Leo Wiener was fluent had also served for this case.

After the stay in Vienna and a brief stay in Holland, the trip to continental Europe was coming to an end, and the family headed to London, where they stayed in a vegetarian hotel in the Maida Vale district, probably following the preferences of Leo Wiener. Norbert recalls visiting Israel Zangwill (1864–1926), a famous English writer with whom his father had corresponded on Zionist issues. Indeed, Zangwill was one of the best known British Zionists, and his is the phrase, "Palestine is a country without a people; the Jews are a people without a country." Wiener would visit Zangwill (whom he described as "interesting" and "sensitive") later, on his second visit to Europe, at the age of 18.

Another character that Leo Wiener was interested in seeing was Prince Pyotr Kropotkin (1842–1921), a noble Russian geographer and writer who later converted to anarchism and was a pioneer in the defense and dissemination of anarcho-communism. It appears that Kropotkin participated in an assassination attempt of his cousin, Tsar Alexander II, and had been forced to leave his native Russia. Interestingly, 30 years later, a son of Kropotkin's secretary would be a student of Wiener.

In September 1901, the European trip came to an end, and the family returned home to Cambridge. The experience had been profitable for a young Wiener who, years later, would write about this trip:

> By the time of my second visit to Europe, when I was a young man, the memory of my first visit, my studies and readings, together with the continued presence of my father, had made Europe nearly as familiar to me as the United States.

Wiener Goes to School

After returning to the United States, Wiener was enrolled in the third grade at Peabody School, a public school in Cambridge. The family moved to Cambridge to a house close to that of M. Bôcher (1867–1918), a mathematician and professor at Harvard since 1894. While Wiener played with Bôcher's children, he could not imagine that his neighbor was a famous mathematician (author, among others, of an important harmonic analysis theorem and several textbooks that were used in the United States), who would die relatively young and in whose honor the American Mathematical Society would create an award in 1923 with its name, the Bôcher Memorial Prize. This prestigious award is currently the oldest awarded by that institution and would be awarded to Wiener himself in 1933, in its fourth edition, due to his contributions in analysis.

Alternating school causes him a few problems, mainly because his education at home had advanced him in some areas, but not in others. After a short time, Wiener's teachers agreed that he would perform better in a more advanced course, and he was sent to fourth grade. Despite this, he did not fit clearly in any grade in the school, losing the sympathy of the teachers who saw in him a very young student with extensive knowledge in certain subjects, but with specific gaps in some areas not expected from the students at that stage of their education. He writes about this period:

> My chief deficiency was arithmetic. Here, my understanding was far beyond my manipulation, which was definitely poor. My father saw quite correctly that one of my chief difficulties was that manipulative drill bored me. He decided to take me out of school and put me on algebra instead of arithmetic, with the purpose of offering a greater challenge and stimulus to my imagination.

Sure enough, after rapid advancement to fourth grade, he was pulled out of school until he entered high school two years later. Meanwhile, his second sister, Bertha, was born in 1902.

His father was solely responsible for his education until 1903. Wiener remembers those years as a tough time, not devoid of severe discipline and moments of reproach. Norbert says that his father always started classes in a friendly, pleasant way until he made his first mistake. Then the loving father would transform into an "angry demon" and he would end up crying. Not even the mother could mediate to defend him.

Regarding the readings of the time, and in addition to the mathematics books, which were already common in his study, Wiener had the opportunity to approach Chinese dictionaries, books on the excavations of Troy, articles on psychiatry, on electrical experiments, or just travel books. In addition, he got two volumes of J. S. Kingsley's *Natural History*, which he would describe as "excellent." Years later,

Kingsley himself would be his professor at the university, which would reinforce his great interest in zoology. He also got in his hands a treatise on electricity where among other things, a primitive "theory of television" was explained (television would not be developed until 25 years later).

Perhaps due to the curiosity aroused by the two natural histories read, Wiener began to hatch the idea of becoming a naturalist. He was fascinated not only by living beings but also at the same time by their complex organization, and writing about this fact, he would say the following:

> Even in zoology and botany, it was the diagrams of complicated structure and the problems about growth and organization which excited my interest.

Curiously, one of his readings at this time was an article on the progress of electrical impulses in the nerves of living beings, which created a certain concern: "the need to conceive an almost alive automaton." Wiener joked about it:

> Cybernetics had a rapid birth [in me]!

In effect, the article dealt with a subject that he would be interested in again years later, with the development of that new discipline called "cybernetics," which he himself would baptize.

Finally, a serious episode marks the life of Wiener: he develops an increasingly growing myopia that progressively prevents him from reading. After the doctor's advice, he had to spend six months without touching a single book, a period of which he says "was probably one of the most valuable disciplines through which I have ever gone" due to the need to study mathematics head-on and languages only by ear. Wiener would wear glasses for life.

Ayer's High School

In the fall of 1903, Wiener returned to school. In this case, his father's intention was to enroll him in high school. This meant that the nine-year-old Wiener would share a seat in class with classmates at least four years older than him. On the other hand, a few months earlier, the family had moved again, this time to a farmhouse near Harvard that Wiener referred to as "the Old-Mill farm." Indeed, it was a farm with an old mill from the 17th century.

Living in the country, caring for a farm that was in full swing (with cows, horses, etc.), teaching at Harvard, translating all 24 volumes of *Tolstoy's Complete Works*, and taking care of Norbert's education, was too much work for Wiener's father. This was the main reason why he chose to return him to school, not without taking the precaution of hearing lessons from the son every night, to be aware of his evolution.

The daring didactic experiment was allowed by Ayer High School, a high school in the small township of Ayer, in Middlesex County, Massachusetts. The place was very well situated for the father because he had to stop there precisely every morning when he went to catch the train to Cambridge. Quickly, and due to his high level of preparation, Wiener was transferred to the third year. It seemed that the delay in his entry into school (except for the brief period in primary school) had not impaired his knowledge but rather the opposite.

Ayer was a small town where Wiener found friends for whom he always had a particular affection. Perhaps because, as he called himself, he was a "socially undeveloped" child, and Ayer was an experience that cultivated him more in the development of the personality than the intellect. Despite the age difference with his classmates, he also made younger friends, some of them younger siblings of his classmates. Wiener tells:

> I owe a great deal to my Ayer friends. I was given a chance to go through some of the gawkiest stages of growing up in an atmosphere of sympathy and understanding. In a larger school, it might have been much harder to come to this understanding. My individuality and my privacy were respected by my teachers, my playmates, and my older schoolmates.

And also:

> I was prepared and ripened for outer world, and for my college experiences.

Indeed, the details of his autobiography about this time are limited to anecdotal adventures and are not very verbose on academic matters. Ayer's school adventure was more of "opening up to the outside world" than finishing high school. Now, as he would later recognize:

Norbert Wiener,
when he was at
Ayer High School.
Courtesy: MIT Museum

My training and social contacts at high school were only the obverse of the coin. The reverse was my continuing recitation to my father at home. My routine when I was in high school scarcely differed from that which had obtained when he was my full teacher. Whatever the school subject was, I had to recite it before him. He was busy with his translation of Tolstoy and could scarcely devote full attention to me even during my recitation hours. Thus I would come into the room and sit down before a father dashing off translations on the typewriter — an old Blickensderfer with an interchangeable type wheel that permitted my father to write in many languages — or immersed in the correction of endless ribbons of galley proof. I would recite my lessons to him with scarcely a sign that he was listening to me. And in fact, he was listening with only half an ear. But that was fully adequate to catch any mistake of mine, and there were always mistakes. My father had reproved me for these when I was a boy of seven or eight, and going to high school made no difference in the matter whatever. Although any success generally led at most to a perfunctory half-unconscious word of praise such as "All right" or "Very good, you can go and play now," failure was punished, if not by blows, by words that were not very far from blows.

At the school, two public debates were organized each month between the pupils, and Wiener decided to participate in one of them for his second year. This prompted him to write that summer a short essay that he titled *The Theory of Ignorance* in which he defended the enforced incompleteness of human knowledge. The piece was to the taste of the father and, even though Wiener himself would later see in it an "inappropriate article" and "far from his reach, given his age," the truth is that this work is mentioned in the different biographies of Wiener as an "antecedent" to the works of the famous logician, Kurt Gödel, who would demonstrate in 1931 the incompleteness of number theory, once its consistency is assumed. Obviously, such a hypothesis is far-fetched since Gödel could not be influenced by the Wiener boy's article, nor was Wiener actually giving any argument that was even remotely similar to Gödel's work.

Finally, only three years after his admission and shortly after his brother Fritz was born, in 1906, Wiener graduated from Ayer at only 11 years old (while his classmates were already 18). The road to college was now a reality.

Tufts College (1906–1909)

The next step in Wiener's academic training was his admission to Tufts College (now known as Tufts University). This stage would end three years later, in 1909, with a "Bachelor of Arts" in Philosophy, with an honorable mention in Mathematics a year earlier than expected.

Thus, on May 8, 1909, shortly before his graduation, a 14-year-old Wiener was featured in the *New York Times* with the headline "BOY OF 14 COLLEGE GRADUATE; Norbert Wiener Soon to Finish at Tufts — Took Course in Three Years," noting that he was known to his peers as "the brightest boy in the world." While Leo Wiener was already known for being a professor at Harvard, a young Norbert Wiener began to become in a certain way the media character on whom all kinds of expectations were poured.

It all started in 1906, before entering the university, when Leo Wiener did not want to risk too much by sending a boy of only 11 years old to Harvard, where the selection of candidates is quite strict. Consequently, he decided to enroll him at Tufts College, where the admission was easier. It was a relatively modern university that had been founded in 1852 by Charles Tufts, an American businessman who donated land for the construction of the campus.

After going through the tests without too many problems and based on his high marks in previous stages, Tufts decided to admit him to the institution. Wiener, who titles the chapter in his autobiography referring to this period "College Man in Short Trousers," thus reflects a much more difficult period than high school. As he wrote:

> I was socially dependent far more on those of about my age
> than on the college students with whom I was to study.

And:

> I was a child of eleven when I entered and was in short trousers
> at that. My life was sharply divided between the sphere of the
> student and that of the child.

At this time, a major change occurs in Wiener's life. While during
secondary education he had been, in some way, privileged by his
young age, in this case the pattern was not repeated, so he found a
more uniform relationship in which he was treated in the same way
as his companions.

The entrance to Tufts College also coincides with another new
move of the family to a house near the university. They placed the
new home in the Medford Hillside district in Medford. Thus, Wiener
would be close to his place of study, as would his father.

Although Wiener's chosen branch was mathematics, he also
showed an interest in languages, presumably influenced by his father.
Thus, he took advantage of this stage to train in Greek and German.
It must be taken into account that the study plans for the degrees
of the time were designed in a very generalist way. There was no
"degree in mathematics" similar to the one we find nowadays both
in the United States and in Latin America or Europe. What Wiener
studied at Tufts is something akin to a degree in philosophy, which
includes, of course, the study of classical and modern languages, with
a specialization in mathematics. The following figure shows Wiener's
academic record. It is curious, for example, to see that many years
after his passage through Tufts, in 1924, his admission to the Phi
Beta Kappa student society as well as his death on March 18 from
1964 were noted in the file.

Regarding the subjects related to his specialty, Wiener fondly
remembered his passage through the course of "Equations Theory"
by a young professor named W. R. Ransom, who would later be the
author (in 1955) of the famous book, *One Hundred Mathematical
Curiosities*. Ransom was probably the only one who understood
Wiener's specific needs, preparing a course for him that would allow
him to delve into more advanced topics, such as Galois theory.
Ransom was a well-known "problem solver" who used to publish
in the Elementary Problems section of the *American Mathematical
Monthly*.

Academic expedient of Norbert Wiener at Tufts College.
Courtesy: Tufts University.

The Problem Solvers

One of the many ways of practicing the profession of a mathematician is that of the problemist (or problem solver), who is concerned with asking and solving curious questions whose interest is probably not comparable to that of a complete theory, but enough to pay attention. Usually, solving this type of problems requires the intelligent use of known results of elementary mathematics (inequalities, series expansions of special functions, groups, linear algebra, synthetic geometry, discrete mathematics, topology, etc.). In mathematical magazines whose spectrum of readers includes undergraduate students, there is usually a problem section. Moreover, there are currently numerous mathematical competitions whose objective is to promote mathematical vocations and early detect mathematical talent in students' youths. These competitions typically consist of solving a small list of problems. There are many books dedicated to their preparation.

In addition to the typical problem solvers that we have just described (among whom we could place Professor Ransom), other types of mathematicians contribute to proposing problems, even offering money to those who solve them. These problems are usually related in some way to situations in the regular investigation of the mathematician who proposes the question. A very famous case is that of Paul Erdös (1913–1996), a Hungarian mathematician who traveled around the world visiting colleagues, staying with them, and contributing to numerous (about 1,500) scholarly articles of great quality. Erdös offered varying amounts of money for the solution of those questions that interested him, but which he did not know how to approach. Many times he presented the award.

Wiener, who argued that he was already "beyond the normal freshman work in mathematics" (somehow highlighting the mediocrity of his classmates and perhaps also the content of the courses), nevertheless remembers the difficulty of this course as the one that demanded the most of him for his entire stay at Tufts. The paths of Wiener and Ransom would later cross again, in 1959, during a visit to Tufts.

Despite his young age and distance from his classmates, Wiener found friendship (and understanding) in Elliot Quincy Adams (1888–1971), a chemical engineering student at MIT and his neighbor. Wiener says that Adams, six years older than him, told him about the possibility of representing four-dimensional figures on a plane and in three-dimensional space. Years later, Adams would graduate from MIT under the famous chemist Gilbert N. Lewis (1875–1946). Lewis stated that "the two most profound scientific minds, among the people I have ever met, were those of Elliot Adams and Albert Einstein." It was precisely on this topic of "representations" that Adams made his most important contribution in 1942:

the work *Planes X–Z in the I.C.I. 1931 Colorimetry*, a seminal contribution to color theory. Wiener would also work with Adams on the development of two electrical appliances. The first was a new electrostatic transformer. They used a series of crystal discs and electrodes connected in parallel to charge it and electrodes connected in series to discharge it. The second device was an electromagnetic coherer (at the time one of the fundamental technological elements for the construction of the radio receiver), different from the model proposed by Edouard Branly (1844–1940) at the end of the 19th century. Although in neither of these two works they obtained more results than "a lot of broken dishes" in the first case and "a device whose operation we were not sure of" in the second, they are Wiener's first contacts with electrical technologies, which also sparked an interest in wireless technologies, an interest that he would keep his whole life. Wiener also remarks on his attraction at that time to chemistry (especially organic chemistry) and physics. Above all, he liked demonstrations, which were a cause of enjoyment for him. He also begins to show a growing interest, which would soon turn into a passion, for biology. In his autobiography, he confesses: "I found [in] the biology museum and [its] laboratory a place of fascination." When talking about his visits to the museum, Wiener also points out that the caretaker and manager of the animal facility had become one of his private colleagues. Thus, this mysterious 12-year-old boy full of curiosity had found better friendship in a person loaded with popular wisdom and stories about the biology museum than in his classmates. His fascination with nature was growing. Although he had already been in Middlesex Falls with a group of students hunting frogs and collecting algae and other living things under the supervision of Professor Fred D. Lambert, this was the moment when a true "biological concern" was born in him. This is why Norbert's father came to consider the possibility that Wiener chose biology as a specialty in future studies. Thus, Leo took his son to Woods Hole, a small town near Falmouth, to visit Professor Parker at Harvard's biological facility. This allowed the young man to carry out dissections of some fish, although according to Wiener, "they were not particularly brilliant" and ended up with the following note on the table where he did them: "Do not dissect fish here." During his last year at Tufts, Wiener decided to take the subject "Comparative Anatomy of Vertebrates" taught by Professor Kingsley, a recognized expert in the field. Wiener was fascinated by Kingsley, whom he describes as a small man, always alert, and to whom he also dedicates the following honor:

> [Kingsley] was the most inspiring scientist whom I met in my undergraduate days.

He also remembers that he spent a lot of time in the laboratory library, surrounded by multiple books on the subject.

J. S. Kingsley (1854–1929)

He was a professor of zoology at Tufts College from 1892 to 1912. He is considered primarily responsible for developing the subject of "comparative anatomy" in the United States in the early 20th century. Through his research, his books (among which is the *Natural History* that Wiener read as a child), and his translations from German into English of several university texts, he was seen in the United States as one of the promoters of this field, currently recognized as one of the main areas of knowledge for any medical student.

Regarding this new vocation, he remarks an inevitable confusion about the reasons that drove him:

> Biological study may have a morbid attraction for a young student. His legitimate curiosity is mixed with a prurient interest in the painful and the disgusting. I was aware of this confusion in my own motives.

Now, there was an incident that caused some tension. It happened in the following way: Wiener and several colleagues practiced, without the knowledge of Professor Kingsley, a vivisection of a guinea pig to perform a certain ligation of arteries. The experiment was a failure, and the guinea pig died. Later, when he heard about it, Kingsley considered it "criminal" and took away from the authors certain privileges over the use of the laboratory. For Wiener, the humiliation itself, and the subsequent guilt, turned out to be more problematic than the punishment itself.

Despite his dabbling in biology, Wiener continued his studies in subjects more typical of mathematics such as calculus and differential equations, still assisted by his father in what he called the "routine of double recitation." Finally, and only 14 years old, Wiener graduated in mathematics in 1909, obtaining the qualification of *cum laude* and completing his studies in one year less than expected. However, he downplayed that it took him less time than his classmates and attributed it to the fact that "only the child can devote his whole life to uninterrupted study."

The Unfinished Adventure in Zoology

Wiener doubly suffered from the customary "horror of emptiness" that often grips many students after graduation. In his case, not only did the doubt arise as to what was the next step to take, but he also became aware for the first time that he was not considered "normal" by his environment. He was a child prodigy and, as such, seemed destined for almost inevitable failure.

What is worse, his father gave several interviews to the press. During them, he made quite clear that, in his view, Norbert was "a normal boy" who "had undergone special training" by him and, therefore, his academic successes were not due to his exceptional skills but simply to the success of the pedagogical methods used. Wiener admits that learning the opinion of his father hurt him. Furthermore, it seems that his father sought his own recognition through the successes of his son. Even so, he was reluctant to believe that everything the father said was his genuine opinion, so, when referring to these facts, he comments:

> I suppose that this was in part to prevent me from being conceited and that it was no more than a half-representation of my father's true belief.

But the damage was done, so he adds:

> Besides the direct damage to me, these articles could only have accentuated that feeling of isolation forced on the prodigy by the hostility latent in the community around him.

Presumably, Wiener would have an inner desire to rebel against the parental authority that was doing so much damage to him at that moment. Perhaps because of this, but also because his contacts with biology had marked him deeply during his undergraduate degree at

Tufts (despite having cursed a degree in mathematics) or, who knows, probably due to the innate curiosity of the adolescent, eager to know the why of life, or motivated by the books read during his childhood, and also by Kingsley's classes, the fact is that he decided to abandon mathematics for zoology. His intention, then, was to pursue a doctorate in zoology. Obviously, this meant breaking with parental wishes for him to study mathematics, but it is also true that those wishes had already been satisfied with his recent graduation.

Taking advantage of the better financial situation that the Harvard professors found after the election of a new president, the Wiener family moved to a better residence. This time the new destination was Cambridge, much closer to Harvard University, where father and son (the former as a professor, the latter as a doctoral student) were to spend much of their time. The decision involved selling the Old-Mill farm and the Medford Hillside house. A time of change was beginning for everyone.

The years between 1909 and 1911 were difficult for Wiener. People were suspicious of a 15-year-old prodigy preparing for his doctorate, while he felt slighted and misunderstood by the adult community. Furthermore, precisely then, he went through a time in which the transitory nature of life began to terrify him, and the fear of death seized him. This fact is curious from the perspective of the years because he would adopt a very different attitude towards death and life itself. When he was older, he would suffer several heart attacks and would be forced to undergo various medical treatments. So, his attitude used to be to see death as inevitable and his body as a good place to experiment. But at 14, Wiener could not have that attitude but one much closer to that of boys of his age:

> Like many other adolescents, I walked in a dark tunnel of which I could not see the issue, nor did I know whether there was any. I did not emerge from this tunnel until I was nearly nineteen years old and had begun my studies at Cambridge University. My depression of the summer of 1909 did not suddenly end; rather it petered out.

The first sign of failure occurred just before entering Harvard when he was denied election as a member of the elitist Phi Beta Kappa society because there was doubt of deserving such an honor due to the uncertainty of the future of a child prodigy. However, the same institution would admit him as a member 15 years later in 1924. This rejection was very significant for Wiener and, in fact, marked his position concerning public recognition, making him very critical of it.

> I have indeed taken a very dim view of all honor societies. This is, of course, the result of my own experience at Tufts, but it has been fortified and strengthened by my subsequent contact with such societies. The fundamental difficulty is that the recognition given by such societies — and, indeed, by universities in the awarding of honorary degrees — is secondary. They do not seek out young men deserving recognition, but award recognition largely on a basis of past recognition. There is, thus, a pyramid of honors for those who already have honors and, per contra, an undervaluing of those who have behind them accomplishments rather than prior recognition.

As if that were not enough for Wiener, his poor visual acuity, his physical clumsiness, and his lack of skills in the laboratory made him realize that the laboratory might not be his place. Indeed, the first classes were somewhat probative of this fact:

> My histology course began as a muddle and continued as a failure. I had neither the manual skill for the fine manipulation of delicate tissues, nor the sense of order which is necessary for the proper performance of any intricate routine. I broke glass, bungled my section cutting, and could not follow the meticulous order of killing and fixing, staining, soaking and sectioning, which a competent histologist must master. I became a nuisance to my classmates and to myself.

By the end of the first trimester, he realized that he could not continue in biology. This, coupled with his inability to maintain the patience necessary for meticulous work, caused him to stop studying what he had longed for during his years at Tufts, and he faced an uncertain future for the first time in his life.

Returning to Mathematics at Cornell

After the year he had somehow lost at Harvard, Wiener won a scholarship to Cornell University, in this case, to return to mathematics and philosophy. Without reaching the prestige of Harvard, Cornell was a major university founded after the Civil War and located in Ithaca in New York State.

Around this time, listening to an after-dinner conversation between his father and Frank Thilly (1865–1934), an ethics professor at Cornell University who was an old acquaintance of the family, Wiener learned that, according to family tradition, they were descendants of the philosopher Maimonides. Curiously, in this haphazard (and belated) way, Wiener first became aware that he was a Jew. Probably due to family indifference to religion, Wiener had never considered his origins up to that point. His mother was also, paradoxically, quite anti-Semitic. Wiener wrote, dramatizing this event to the maximum:

> I looked in the mirror and there was no mistake: the bulging myopic eyes, the sightly everted nostrils, the dark, wavy hair, the thick lips. They were all there, the marks of the Armenoid type.

Not only did he not find his discovery interesting, but also he added it to the list of misfortunes and worries that then fueled his insecurity and his depression. What could have started as an exciting time in his life, a "return to the ancestors," trying to take the example of Maimonides and study philosophy with enthusiasm became a heavy burden for him.

Regarding the studies at Cornell, things did not go well at all. Wiener took a course called Complex Variable Function Theory, taught by one Hutchinson. He himself would comment on the course's difficulty that "it was beyond me." Although Wiener attributes it to his immaturity, he also finds the cause that the course did not have a good approach to the logical difficulties of the subject. It would not be until a few years later that, at the hands of the English mathematician G. H. Hardy, Wiener would return to similar topics.

As the year progressed, he became more and more aware that his scholarship was not going to be renewed, and this is precisely what happened.

With a deep sense of guilt for having failed his family, Wiener was transferred by his father to Harvard, to the Philosophy Department. Once again, the meddling of Leo Wiener in the life of his son, a graduate in the body of a teenager, would have the effect of lowering his self-esteem.

Back to Harvard

In 1911, Wiener returned to Harvard to obtain a doctorate in philosophy. As he recounts, the period he spent there coincided with the stay in the department of several great names in philosophy. The first of them who was his teacher, and had a positive influence on him, was Jorge Santayana (1863–1952), a naturalist philosopher born in Madrid of a Spanish mother and settled in the United States from a very young age (although he never accepted the American citizenship). Other important philosophers with whom he had some contact were J. Royce (1855–1916) (of whom he was a student for two courses, one on mathematical logic and the other on the scientific method), G. H. Palmer (1843–1933), and H. Münsterberg (1863–1916) (of whom he does not say much and, apparently, who did not have a remarkable influence on his formation).

Portrait of J. Santayana (1863–1952). Samuel Johnson Woolf (1880–1948). Courtesy: Time magazine, Public domain, Wikimedia Commons.

Wiener comments that, despite not retaining a clear idea about the content of Santayana's classes, he does remember perfectly how the feeling of continuity with ancient culture and how philosophy was an intrinsic part of life filled him with satisfaction. The question is compelling: Was Santayana the first professor to make this impression on Wiener, which was a feeling that had not appeared for him since his years at Tufts College? Perhaps the luck of finding someone so inspiring was the necessary support to gather strength and, finally, do two good years at Harvard that ended with his well-deserved award, the doctorate in philosophy.

Regarding Royce, Wiener explains that although his contributions to mathematical logic cannot be considered of importance, in his lectures it became clear that you were facing a very intelligent person, who perhaps had not achieved the advances he deserved due to the simple fact of having arrived too late to the subject. In addition, the truth is that it was he who introduced Wiener to the world of logic and, what is more, Royce is currently considered the founder of the Harvard School of Logic, Boolean Algebras, and Philosophy of Mathematics. On the other hand, Wiener learned the purely mathematical aspects, which he would need to develop his doctoral thesis, in another course of mathematical logic from the hands of E. V. Huntington, an old friend of his father.

Edward V. Huntington (1874–1952)

Edward V. Huntington
(1874–1952).
Courtesy: Public domain,
Wikimedia Commons.

Born in Clinton, New York, on April 26, 1874. He graduated from Harvard in 1895, in a class where he had as partners J. L. Coolidge and J. K. Whittemore. Two years later, he became a math instructor at Williams College. From 1899 to 1901, he was in Europe finishing his doctorate in Strasbourg on the fundamentals of number systems, foreshadowing the topic to which he would later dedicate his best works on axiomatics. Following this, he returned to Harvard, where he would remain as a professor until his retirement in 1941.

Huntington gave sets of axioms for many mathematical systems, showing particular emphasis on the independence of each set of postulates that he considered, giving examples of mathematical objects that satisfied all but one of the axioms. He worked on different axiomatics for groups, abelian groups, Boolean algebras, plane and spatial geometry, and the fields of real and complex numbers.

In 1904, he proposed an axiomatization of Boolean algebras. This axiomatization would be revised again in 1933, proving that a Boolean algebra requires only a commutative and associative binary operation (denoted below by "+"), and a unary operation, the complement, denoted below with a prime. In this way, he proved that the only additional axiom that a Boolean algebra requires is

$$(a' + b')' + (a' + b)' = a,$$

which is currently known as the "Huntington Axiom."

In another vein, he proposed a method for the apportionment of the treasury in the United States House of Representatives, which was approved in 1941 and is still in use. He also made some contributions to electoral mathematics.

In 1919, Huntington was the first president of the Mathematical Association of America (an association that he helped to found), and in 1924, he was vice president of the American Mathematical Society (AMS).

On Huntington, Wiener would comment the following:

> (...) He was a magnificent teacher and a very kind man. His exercises in postulate theory were all educational gems. He would take a simple mathematical structure and write a series of postulates for which we were to find not only examples satisfying the complete list, but other examples failing to satisfy it in just one place or in several specified places. We were also encouraged to draw up sets of postulates of our own. (...)
>
> Huntington's career has always remained a mystery to me. With his keenness and his inventiveness I should have expected some great mathematical contribution from his pen. Nevertheless, all his pieces of work, no matter how much they may contain of ideas, have remained miniatures and vignettes (...) The valuable and honorable career of Huntington seems to me to bear the lesson that one of the most serious possible lacks in mathematical productivity is the lack of ambition, [and I think] that Huntington simply set his sights too low.

Wiener decided, for his doctoral thesis, to tackle a technical question of the new mathematical logic. Specifically, he conducted a comparative study between Schröder's logic of relations and the more recent one by Russell and Whitehead. In principle, his advisor was Royce, but he fell ill and could not continue. However, Wiener had the good fortune that K. Smidth, a young Tufts College professor and also an expert in mathematical logic, agreed to replace Royce. Finally, both signed as directors of the thesis.

To complete the doctorate, he had to pass a series of exams, first written and then oral. The written examinations did not seem to worry him, and he took them without great difficulty. However, when it came time to pass the oral exams, attending the corresponding teacher's home for each examination and submitting to his questions, Wiener felt paralyzed with fear:

> I must give my father every credit of seeing me through the great ordeal of the oral examinations. Every morning he went for a walk with me to keep up my physical condition and to reinforce my courage. Together we walked over many parts of Cambridge which were as yet unknown to me. He would ask me questions concerning the examinations that were ahead and would see to it that I had a fair idea as to how to answer them.
>
> Nevertheless, at my own valuation, I should have failed every examination; but examining professors for the doctorate are likely to be more human and sympathetic with the student than the student is with himself, and to give him the benefit of every doubt.

Norbert Wiener, Ph.D. when he was 18 years old, 1913. Courtesy: MIT Museum.

Once the oral exams had been passed, the next and last step was the presentation of the doctoral dissertation itself before the team of professors from the Harvard philosophy department. For this examination, Wiener had no difficulties because, as he points out in his autobiography,

> (...) no well-regulated department will permit a candidate to proceed so far unless it is substantially sure that he is going to pass.

So, at the age of 18, Wiener earned his Ph.D. degree from Harvard, something his father could be proud of. His didactic experiment had paid off.

Chapter 2

Beginnings as a Mathematician

Wiener Abandons Philosophy

After defending his doctoral thesis, Wiener got a Harvard University scholarship to travel to Europe and continue his studies. Taking advantage of this circumstance, his father took a sabbatical year intending to have the whole family visit the old continent and, probably, with the idea of closely monitoring Norbert. Furthermore, Leo Wiener wanted to earn the respect of European teachers, and both he and his wife wanted his daughters to acquire at least a veneer of European culture. Although the family's base of operations would be in Munich, Leo accompanied his son to Cambridge, where they contacted Bertrand Russell, who would act as Wiener's tutor.

From the very beginning, the relationship between Russell and Wiener was tense. Russell did not like "prodigies." On the contrary, Wiener failed to impress him and caused mistrust by not showing the discursive capacity essential for every philosopher. Wiener met a man he would soon describe (in correspondence with his father) as an iceberg because of his excessive coldness.

In fact, within days of his father's departure, Wiener wrote him a letter in which he complained about the treatment he received from Russell when he visited him at tutorial sessions:

> Russell's attitude seems to be a mixture of contempt and complete indifference. I think I will have to settle for what I get out of him in his classes ...

Wiener attended two courses taught by Russell, one on *Principia Mathematica* and the other on "Sense and Perception." Nevertheless, the experience was rather negative. When he had been attending his classes for about a month, Wiener wrote to his father again. This time, he expressed his complete dissatisfaction and confessed to having severe doubts about his capacity as a philosopher. Russell had given him a strong harangue in which he accused him of being over-confident and, to top it all, called his philosophical views a "horrible fog" and said that the exposition he had made of them was even worse. Wiener's reaction was immediate. He went from admiring the philosopher to despising the man.

However, in his autobiography, Wiener would fondly recall his early experiences at Cambridge:

> (...) This was the first time that I had come to be really competent to live alone and could learn something about the freedom of an independent worker.

Furthermore, instead of complaining about the treatment received from Russell and, undoubtedly, thanks to the vision that the perspective of the years gives us, he appreciated the advice received:

> Russell impressed upon me that to do competent work in the philosophy of mathematics, I should know more than I did about mathematics itself.
>
> Hardy, to whom I turned, was an ideal mentor and model for an ambitious young mathematician (...)

And, a few pages later:

> At Cambridge, Russell had impressed on me not only the importance of mathematics but the need for a physical sense, and he had suggested that I study the new developments of Rutherford and others concerning the theory of the electron and the nature of matter.

In contrast to his experience with Russell, Wiener would find in G. H. Hardy (1877–1947) comfort and inspiration:

> His course was a delight to me. My previous adventures into higher mathematics had not been completely satisfying, because I sensed gaps in many of the proofs which I was unwilling to disregard — and correctly too, as it later turned out, for the gaps were really there and they should have disturbed not only me but my earlier teachers. Hardy, however, led me

through the complicated logic of higher mathematics with such clarity and in such detail that he resolved these difficulties as we came to them and gave me a real sense of what is necessary for a mathematical proof. He also introduced me to the Lebesgue integral, which [is the tool that] was to lead directly to the main achievements of my early career [as a mathematician].

Indeed, it can be said that Wiener rediscovered mathematics under the guidance of Hardy and, on the other hand, following Russell's advice, cultivated an excellent predisposition towards physics and engineering, which he perfectly combined with always keeping a watchful eye on applications outside of mathematics. This would also be of great help in his professional life since most — if not all — of his academic career was developed in an institution dedicated almost exclusively to engineering research, the Massachusetts Institute of Technology (MIT).

It might seem strange that Wiener, having already defended a thesis on the philosophy of mathematics, needed help to understand the Lebesgue integral. But, as it happens, mathematics is a particular discipline. If we compare it with physics, chemistry, or any other scientific field, we quickly see something different. For example, the nature of experiments in mathematics is entirely different from those in other disciplines. In mathematics, only so-called thought experiments are possible, which consist of imagining that certain circumstances exist and calibrating the consequences that we would draw from them with the sole force of reasoning and the only limits of logic. This means that in mathematics, it is impossible to let the laws of nature do their work, observe what happens, and, a posteriori, look for an explanation, as is usual in physics.

To be honest, this is no longer quite the case because the computation capacity that we currently enjoy allows us to program certain types of situations, carry out simulations, and look for an explanation with the results before our eyes. But this was not the case at the beginning of the 20th century. This specific characteristic of experiments in mathematics makes this discipline especially difficult for many people since one must do all the work from the beginning and also in a way that is simultaneously rigid and requires great imagination. In particular, this demands a lot from the beginner. In addition to discipline, a personal relationship must be established with the basic concepts and results. For this relationship to mature and bear fruit, intellectual work is not enough. It takes a certain amount of time. It is usual that when faced with a new theory, we can follow each step and, therefore, we see the truth of the results, and this gives us the feeling of having understood things. However, in many of these cases, we have not understood even ten percent of what we are told. What are the ultimate reasons why this or that has been tried? Why do certain things seem more natural than others? Why does something that was initially easy and intuitive become difficult and improbable? Every mathematician has had these kinds of experiences. Therefore, as was the case with Wiener, it is natural that someone who had studied mathematics in a hurry and at a very young age needed the guidance of another to assimilate in full some of the most sophisticated theories of the time. And such was the case with the Lebesgue integral. It is also natural that he was grateful to Hardy all his life since there is no doubt that the road would have been very hard without his help.

G. H. Hardy (1877–1947), when he was 23 years old. Courtesy: Oberwolfach, Public domain.

Hardy was a true genius whom Wiener, in his mature days, considered the most important English mathematician of his generation. From the very beginning, they became good friends, and, in time, Hardy would become one of the mathematicians who published Wiener's work in the United States. With Hardy, Wiener would study the Lebesgue integral and the foundations of the theory of functions of one complex variable. Both subjects would later be fundamental to the proper development of Wiener's research.

Thus, expressed in symbolic terms, Hardy's positive influence, guiding Wiener towards mathematical research, was coupled with Russell's apparent contempt for Wiener as a philosopher, a disdain which he took as a fulcrum. Leaning on it, he was driven by a powerful force towards mathematics, gradually abandoning his penchant for philosophy. However, he still had time to make some contributions to the philosophy of mathematics, in particular, to the study of various axiomatic systems. For example, in 1920, he published an article in the *Transactions of the American Mathematical Society* in which, following Huntington's line of thought, he succeeded in establishing a set of axioms that characterize the algebraic field structure, based on a single binary operation, which he denoted by x@y and that, when we use the usual notation based on a product and a sum, shares the formal properties of the operation $1 - x/y$. Moreover, he would also deal with philosophical issues related to the profession and, in particular, with the scientist's responsibilities — but we will talk about this later.

Göttingen

Wiener did not spend the entire academic year at Cambridge. In the spring term, he traveled to Göttingen, where he would attend the classes of D. Hilbert (1864–1943) and E. Landau (1877–1938). Moreover, he took advantage of Russell's advice and studied some physics papers. Particularly interesting were the articles published in 1905 by a young patent inspector named A. Einstein (1879–1955) on three leading topics in physics at the time: the photoelectric effect, the Brownian motion, and the special theory of relativity. Of these articles, we are especially interested in the one dedicated to Brownian motion since it was in this subject that Wiener would later make his first important mathematical contributions.

At the end of the 19th century, physicists and chemists were immersed in an intense controversy on the very nature of matter. What was in dispute was whether, as Democritus had predicted in ancient Greece, the matter is composed or not of certain elementary, indivisible particles, which he called atoms. These particles, it was thought, had to combine following specific rules to form other larger particles, molecules, which are what define the physical and chemical nature of different materials. The most important physicists supporting the atomic hypothesis were J. C. Maxwell (1831–1879) and L. Boltzmann (1844–1906). In particular, Maxwell had developed his kinetic theory of gases based on the fact that they were made up of molecules in perpetual motion and, after establishing that the temperature of these gases was directly linked to the movement of their molecules (the greater the amount of motion, the higher the temperature, and vice versa), had deduced rigorous explanations for some basic properties of these. For his part, Boltzmann derived the basic principles of thermodynamics from the atomic hypothesis. As the 19th century progressed, more and more physicists accepted this hypothesis. However, others like W. Ostwald (1853–1932) and

G. Helm (1851–1923) rejected it outright. For them, the crucial concept in physics was that of energy, and for this reason, they called themselves "energetics." Others, like E. Mach (1838–1916), although they rejected the atomic hypothesis because they did not want to include among their theories the existence of specific elements or particles that they could not visualize and with which they could not interact from a sensory point of view, did admit the pedagogical value of this hypothesis since it led to explanations of numerous phenomena whose interpretation from a purely macroscopic point of view seemed difficult to achieve. The tension became tremendous, and, for some time, the balance was tipped on the energy activists' side, so much so that Boltzmann, in the introduction to the second part of his *Theory of Gases*, wrote:

> It would be a great tragedy for science if the theory were temporarily thrown into oblivion due to a momentary hostile attitude towards it. I realize that I am nothing more than an individual who weakly fights against the currents of the time ...

Some historians of science attribute Boltzmann's later suicide (in 1906) to his extreme pessimism about the future of kinetic gas theory.

In 1905, the balance tipped to the side of the atomists. That year, A. Einstein (1879–1955) and M. Smoluchowski (1872–1917), independently and, in fact, following quite different paths, formulated an explanation of the Brownian motion from the atomic hypothesis. What's more, they used their theories to calculate Avogadro's number. But what is this about Brownian motion?

Suppose we deposit a group of microscopic particles in a fluid such as, for example, water, and observe what happens under the microscope. In that case, we will see that each one of these, instead of gradually sinking, which would be in accordance with the law of gravity, suffers sudden changes in direction and speed, sometimes accelerating and sometimes slowing down, going up or down, and so on. Furthermore, there are two particularly intriguing issues in the observed movement: It is a motion that does not stop and it is absolutely erratic and unpredictable. Where does each particle get the energy needed to make so many jumps, so many pirouettes?

This phenomenon was first observed by the Dutch biologist and merchant, inventor of the microscope, A. van Leeuwenhoek (1632–1723), but it was the Scottish biologist R. Brown (1773–1858) who first proposed, in 1828, an explanation. The particles introduced by Brown under the water were male sex cells from plant pollen, so his first explanation was that the vital movement of the sperm was the real driver of the particles. However, Brown did not take long to discover that the same phenomenon also occurred when submerging other (non-sexual) cells of plants. So, he reformulated his hypothesis, arguing that there must be an elemental molecule of life that should be responsible for the movement that was observed. It became a topic of almost popular interest, as it seemed to be associated with the mysterious phenomenon of life's origin. However, Brown also observed the phenomenon by submerging inanimate particles in water, and then he was left with no answers.

The Einstein–Smoluchowski theory of Brownian motion is founded on a few elementary principles. To explain it and, for simplicity, we focus on a single spatial component of the motion of the particle. That is, we perform the calculations for the one-dimensional Brownian motion as follows.

Let $\int_{-\infty}^{y} p(x|h,t)dh$ denote the probability that a particle that, for the instant $t = 0$ was at position x, be at some point of the interval $(-\infty, y]$ at time t. Then, the probability that a particle that, for the instant $t = 0$ was at position x, be at instants $t_1, t_2, \ldots, t_n$ ($0 < t_1 < t_2 < \cdots < t_n < \infty$) on the intervals $(\alpha_1, \beta_1), (\alpha_2, \beta_2), \ldots, (\alpha_n, \beta_n)$, respectively, is given by:

$$\int_{\alpha_1}^{\beta_1} \int_{\alpha_2}^{\beta_2} \cdots \int_{\alpha_n}^{\beta_n} p(x_0|x_1, t_1) \times p(x_1|x_2, t_2 - t_1) \times \cdots$$

$$\times p(x_{n-1}|x_n, t_n - t_{n-1})dx_1 dx_2 \cdots dx_n.$$

Moreover, for $\Delta t \to 0$, the following asymptotic relationships should hold:

$$\int_{-\infty}^{\infty} (x - x_0)p(x_0|x, \Delta t)dx \sim F(x_0)\Delta t$$

$$\int_{-\infty}^{\infty} (x - x_0)^2 p(x_0|x, \Delta t)dx \sim 2D\Delta t$$

$$\forall k \geq 3 : \int_{-\infty}^{\infty} |x - x_0|^k p(x_0|x, \Delta t)dx = o(\Delta t),$$

where $F(x)$ is the value of the force field that acts on the particle when it is at the point x and D is the diffusion constant, which is given by $D = \frac{2kT}{f}$, where k is the Boltzmann constant, T is the temperature, and f is the coefficient of friction of the liquid in which the particles are immersed.

Well, from the first of the previous identities, it can easily be proved that

$$p(x|y, t + s) = \int_{-\infty}^{\infty} p(x|\xi, t)p(\xi|y, s)d\xi \text{ (all } s > 0),$$

and, from this point on, using the previous asymptotic identities, it is demonstrated that $p(x|y, t)$ satisfies the differential equation

$$\frac{\partial p}{\partial t} = D\frac{\partial^2 p}{\partial x^2} + \frac{\partial}{\partial x}(F(x)p).$$

If we take into account the initial conditions of the problem and also require that there is no intervening external force field (that is, if we assume that $F(x) = 0$), then it is easy to check that the only solution of the above differential equation is given by

$$p(x|y, t) = \frac{1}{2\sqrt{\pi D t}} \exp\left(-\frac{(y - x)^2}{4Dt}\right).$$

In particular, the mean square deviation (with respect to the abscissa x) of the particle at the instant of time t is given by

$$\int_{-\infty}^{\infty} (y - x)^2 p(x|y, t)dy = 2Dt.$$

This means that one could empirically calculate this mean and, from the data, obtain a reasonable approximation of D and, henceforth, of Avogadro's number.

Interestingly, when he wrote his first paper, Einstein was unaware of the observations made by Brown, so he predicted the existence of

Brownian motion directly from the atomistic hypothesis. In his 1905 article, we can read:

> If the movement discussed here can actually be observed (together with the laws relating to it that one would expect to find), then classical thermodynamics can no longer be looked upon as applicable with precision to bodies even of dimensions distinguishable in a microscope: an exact determination of actual atomic dimensions is then possible. On the other hand, had the prediction of this movement proved to be incorrect, a weighty argument would be provided against the molecular-kinetic conception of heat.

The fact that this was a phenomenon familiar to many (and still lacking a reasonable explanation) was therefore one of the strengths of Einstein's theory. Precisely the successful calculation of Avogadro's number from these data, which was carried out between 1908 and 1914 through different experiments by J. B. Perrin (1870–1942) and his students, constituted the final blow that the atomists needed for their consolidation. In fact, Perrin received the Nobel Prize in physics in 1926 for these calculations.

Wiener Comes to MIT

From Göttingen, Wiener returned to the United States at the same time that the Great War broke out in Europe. He spent the summer in New Hampshire and then traveled to England. But, under these circumstances, he did not find the necessary peace of mind to do serious research work, nor did he find anyone who could collaborate with him. In the late winter of 1915, and at his father's request, he returned to the family's home in Boston.

The next four years were tough and uneasy. Wiener, in his autobiography, recalls that

> ... The war took some years to come to America, but it was never out of my thoughts. The present generation, which has been brought up with crisis as a daily associate, grown up with the crisis as an indefatigable associate, can scarcely be aware of the shock with which the war impinged on my contemporaries ...

Norbert Wiener (first from right)
at the Aberdeen Proving Ground,
where he performed ballistic
calculations in 1918, shortly
before joining MIT.
Courtesy: MIT Museum.

In addition, he changed jobs several times. First, he was an assistant professor in the philosophy department at Harvard University, a position he gave up because he was not promoted. Later, he was hired as a mathematics professor at the University of Maine. This job was achieved through the intermediacy of a teacher agency, something that, in his own words, made him feel "humiliated" by not being able to get a position thanks to the prestige of his "academic merits." To top it all, he was overwhelmed there when he was unable to maintain order in his classes. Other jobs he did in this time period included the following: contributor to the Encyclopedia Americana, columnist for a Boston newspaper, soldier, and mathematician involved in calculating ballistic tables at the Aberdeen Practice Ground, Maryland (when the United States finally entered the conflict), among others.

In 1919, with the support of a friend of his father's, Professor W. F. Osgood of the mathematics department at Harvard University, Wiener got a job as a professor at the Massachusetts Institute of Technology (MIT). There he would stay the rest of his life.

Mathematical Interlude: The Lebesgue Integral

Lebesgue's theory of integration had a fundamental impact on Wiener's spirit and scientific training. This theory was born with the intention of becoming an essential branch of pure mathematics, giving a solid foundation to broad extensions of mathematical analysis. In particular, thanks to the concept of Lebesgue integral, it was possible to establish some of the main theorems of Fourier Analysis. However, in the eyes of Wiener, the Lebesgue integral was the necessary tool to clearly and definitively establish numerous concepts of the calculus of probabilities and mathematical statistics (and, what's more, statistical physics), becoming a fundamental subject of applied mathematics.

The first steps towards a general theory of integration had been taken first by A. L. Cauchy (1789–1857) and then by G. F. B. Riemann (1826–1866), in their attempts to resolve some remaining gaps in the study of trigonometric series. Indeed, J. B. J. Fourier (1768–1830), in a series of works related to the problem of the distribution of heat in conductive solids, had introduced an efficient method for the study of the associated differential equations, which consisted of assuming that the solutions are superpositions of relatively simple solutions, which can be expressed as a product of functions of one variable. In the case of the heat equation for a finite rod of length π, this hypothesis translates to the requirement that an arbitrary function $f(x)$, defined for the values $x \in [0, \pi]$, can always be represented as a sum of the type $f(x) = \sum_{k=1}^{\infty} b_k \sin(kx)$. Fourier had managed to show that for certain classes of (very smooth) functions $f(x)$, the previous expression is true as long as the numbers $b_k = \frac{2}{\pi} \int_0^{\pi} f(x) \sin(kx) dx$ are taken as coefficients. This assertion had in fact been proven previously by Euler, and even with a level of rigor much higher than that used by Fourier, but Fourier went further in stating that these series expansions should be valid for arbitrary functions. The basic idea on which he supported his argument was that for the calculation of the coefficients b_k it is not necessary at all to assume any smoothness on the function $f(x)$ (nor that it is given in its entire domain of definition by a single analytic expression, as D'Alembert (1717–1783) had demanded), but the only essential thing is to suppose that the area enclosed by $\frac{2}{\pi} f(x) \sin(kx)$ between the abscissas 0 and π is finite for every k. Thus, the following question remained in the air: What functions $f(x)$ are such that its graph defines a specific area between each pair of abscissas a, b ($a < b$) of its domain?

The first answer came from Cauchy, who introduced the integral $\int_a^b f(x) dx$ as a limit of sums of the type

$$\sum_{i=1}^{n} f(x_i)(x_i - x_{i-1})$$

(where $a = x_0 < x_1 < \cdots < x_n = b$ and we take the limit for $\max_{i \leq n} \{x_i - x_{i-1}\} \to 0$). He showed that if $f(x)$ is continuous on $[a, b]$, this integral is well defined and finite. In particular, this result supports the degree of generality of the functions that can be expanded as a Fourier series since the Fourier coefficients are well defined for every continuous function and therefore all these functions have an associated Fourier series expansion whose convergence is worthy of study. Note that continuity is not incompatible with the same function having different analytical descriptions in different parts of its domain of definition. For its part, J. P. G. L. Dirichlet (1805–1859) argued that continuity is not a prerequisite for the existence of the Cauchy integral, since this integral exists for all functions that only have a finite number of jump discontinuities and, indeed, also exists for some functions with infinitely many discontinuities.

On the other hand, he also introduced an example of a function that is not integrable in Cauchy's sense. Concretely, he considered the function $f(x)$ that is equal to 0 for rational values of the indeterminate x and 1 for irrational values of x. As we can take the partition $a = x_0 < x_1 < \cdots < x_n = b$ arbitrarily, we can force on the one hand that all points $x_i, i = 1, \ldots, n-1$ are rational (in which case, the associated sum is 0) or that they are irrational (in which case, the sum is close to the value $b - a > 0$). It follows that Fourier's statement does not make sense for arbitrary functions, and therefore we have to study in depth for what kind of functions we can define the Fourier coefficients. Dirichlet believed that the necessary condition for the existence of the integral of a function was that the set of points of discontinuity is a scattered set (that is, a set whose closure has an empty interior). However, Riemann would show that this condition is not necessary by giving examples of integrable functions with dense sets of discontinuities!

Riemann's example. Consider the function $h(x) = x - e(x)$, where $e(x)$ denotes the nearest integer to x, with the convention that if $x = n + \frac{1}{2}$ for certain integral number n, then $e(x) = n$. Let us then define $f(x) = \sum_{n=1}^{\infty} \frac{1}{n^2} h(nx)$. This function has a jump discontinuity at every point of the form $\frac{p}{2q}$ with $p, q \in \mathbb{Z}$, p odd, and $g.c.d\,(p, q) = 1$. Obviously, these points form a dense subset of the real line. And yet, Riemann proved that $f(x)$ is integrable on every closed interval $[a, b]$!

Indeed, in 1854, Riemann took for his Habilitation the problem of the representation of general functions as Fourier series. The most important result that had been proven at that time was due to Dirichlet, who in an article published in 1829, had shown that Fourier expansions are convergent for functions with a finite number of jump discontinuities. In order to improve Dirichlet's result, Riemann again had to face the question of which functions can be integrated, and it is in this context that he provided the example just described. In addition, he slightly modified the Cauchy integral concept, by transforming the sums $\sum_{i=1}^{n} f(x_i)(x_i - x_{i-1})$ into sums of the type $\sum_{i=1}^{n} f(t_i)(x_i - x_{i-1})$ with $t_i \in [x_{i-1}, x_i]$ and demonstrated a necessary and sufficient condition for the existence of this new integral (which we do not explain now due to its excessive complexity). The important thing is that he used his characterization of integrability to show that the function given in the previous example is integrable in the Riemannian sense and, therefore, Dirichlet was wrong in his conjecture about the extension of the class of the integrable functions.

In 1892, C. Jordan (1838–1922) introduced the concepts of interior and exterior content of a set of points $A \subset [a, b]$ as follows: Given a partition P of $[a, b]$ in intervals $I_k = [t_{k-1}, t_k]$ (with $a = t_0 < t_1 < \cdots < t_n = b$), we consider the sums $i(P, A) = \sum_{I_k \subseteq A} (t_k - t_{k-1})$ and $e(P, A) = \sum_{I_k \cap A \neq \emptyset} (t_k - t_{k-1})$. Then, the inner content of A is defined by $c_i(A) = \sup_P i(P, A)$ and the outer content by $c_e(A) = \inf_P e(P, A)$. Obviously, $c_i(A) \leq c_e(A)$. We say that A is measurable in Jordan's sense if both values coincide and, in this case, we call this value the content (or the Jordan measure) of A and denote it by $c(A)$. These concepts allow us to redefine the Riemann integral in the following terms: Given a partition $P = \{E_j\}_{j=0}^{\infty}$ of $[a, b]$ formed exclusively by sets E_j that are measurable in Jordan's sense and pairwise disjoint, we introduce, for any bounded function $f(x)$ and every set E_j, the amounts

$$m_j = \inf_{x \in E_j} f(x), \quad M_j = \sup_{x \in E_j} f(x),$$

$$s(f, P) = \sum_{j=0}^{\infty} m_j c(E_j), \quad S(f, P) = \sum_{j=0}^{\infty} M_j c(E_j),$$

and we say that f is integrable on $[a, b]$ when $\inf_P S(f, P) = \sup_P s(f, P)$. Moreover, in such a case, the integral (in Jordan's sense) is given by

$$\int_a^b f = \inf_P S(f, P) = \sup_P s(f, P).$$

This concept, formalized in this form, is already very close to Lebesgue's definition of integral. The only issue to bear in mind is that the partitions that Jordan used to define the content of the sets are finite, while in the case of the Lebesgue measure, we would deal with arbitrary partitions. But the move from one concept to the other was forced because Jordan's content had some strange properties. For example, not all countable sets have the same Jordan content. Furthermore, mathematically speaking, Jordan's integral had some problems with some limit-passing processes, making it an unpleasant object.

The next important step was taken by E. Borel (1871–1956) and, curiously, the motivation did not come from Fourier analysis but from an apparently distant subject: the problem of analytical prolongation of functions of a complex variable. In his doctoral thesis, Borel studied the behavior of certain series of complex variable functions that were known to diverge in a dense subset of a closed curve. Specifically, he was interested in the series of the form $f(z) = \sum_{n=0}^{\infty} \frac{a_n}{z - \alpha_n}$, where $\sum_{n=0}^{\infty} |a_n| < \infty$ and $\{\alpha_n\}_{n=0}^{\infty}$ is assumed to be a dense subset of a certain closed curve C. It was thought that the functions defined by such series could not be analytically continued beyond the curve (in fact, such is the case if we consider Weierstrass's concept of analytic prolongation). However, Borel showed that we could cover the points on the curve where divergence occurs with a succession of curve arcs of length so small that the total sum of these arcs' lengths falls below any positive number fixed in advance. It follows that there are many more points on the curve where convergence does occur than points where it does not, so we can consider curves that cross from one side to the other of the closed curve where the function is well defined and continuous. In this way, we can interpret that the function "is the same" on both sides of the curve.

Motivated by this discovery, Borel introduced in 1898 a very general measure theory for the subsets of the interval $[0, 1]$. He realized that every open subset of the real line can be expressed as a union of a finite or infinite countable family of pairwise disjoint intervals and, therefore, a measure can be assigned to these sets simply by adding the measures of the different intervals. Based on this idea, Borel extended the concept of measure first to the open subsets and then to arbitrary unions of open subsets of $[0, 1]$, and, through the calculation of the complement, he also defined $m(A^c) = 1 - m(A)$. In this way, iterating the process over and over again, we come to define a measure on the family $\mathbf{B}$ of subsets of $[0, 1]$ characterized by the following elementary properties:

(i) All open subset of $[0, 1]$ belongs to $\mathbf{B}$.
(ii) If $\{A_k\}_{k=1}^\infty \subseteq \mathbf{B}$, then $\bigcup_{k=1}^\infty A_k \in \mathbf{B}$.
(iii) If $A \in \mathbf{B}$, then $A^c \in \mathbf{B}$. (The elements of $\mathbf{B}$ are called Borel sets of $[0, 1]$.)

We can now introduce the Lebesgue measure. Suppose that $a < b$ and $A \subset (a, b)$. Obviously, there will be many ways to cover the whole set A with a countable family of intervals $I_i = (a_i, b_i)$ (that is, $A \subseteq \bigcup_{i=0}^\infty I_i$), and it is possible that for some of these families we have that $\sum_{i=0}^\infty (b_i - a_i) < b - a$. Well, we call external measure of A to the quantity

$$m^e(A) = \inf \left\{ \sum_{i=0}^\infty (b_i - a_i) : A \subseteq \bigcup_{i=0}^\infty (a_i, b_i) \right\}.$$

On the other hand, if we take $A^c = \{x \in (a, b) : x \notin A\}$, we can also consider the quantity $m^e(A^c) \leq b - a$. We call the internal measure of A the number $m^i(A) = b - a - m^e(A^c)$ and say that the set of points A is measurable, with measure $m(A)$ if $m(A) = m^i(A) = m^e(A)$. Otherwise, we will say that A is not measurable in the Lebesgue sense. Finally, we will say that the set $A \subseteq \mathbb{R}$ is Lebesgue measurable if for every open interval (a, b) we have that $A \cap (a, b)$ is Lebesgue measurable and, in this case, the measure of A is given by

$$m(A) = \lim_{a \to -\infty; b \to +\infty} m(A \cap (a, b)).$$

This way of "measuring" the size of sets is the best that can be achieved if one tries to introduce a concept of length for the subsets of the real line. In particular, it can be proved that it is impossible to introduce this concept of length in a coherent way without excluding some subsets of the line. Furthermore, in an article that appeared in 1918, W. Sierpinski (1882–1969) showed that the class of subsets of $\mathbb{R}$ measurable in the Lebesgue sense is axiomatically characterized precisely as the broadest possible class Θ for which it is possible to define a function (measure) $\mu : \Theta \to [0, \infty]$ that verifies the following axioms:

(i) If $a < b$, then $[a, b] \in \Theta$ and $\mu([a, b]) = b - a$.

(ii) If $\{A_i\}_{i=1}^{\infty} \subseteq \Theta$ is a finite or infinite countable family of pairwise disjoint elements of Θ, then $\bigcup_{i=1}^{\infty} A_i \in \Theta$ and

$$\mu\left(\bigcup_{i=1}^{\infty} A_i\right) = \sum_{i=1}^{\infty} \mu(A_i).$$

(iii) If $A_1, A_2 \in \Theta$ and $A_1 \supset A_2$, then

$$A_1 \setminus A_2 = \{a \in A_1 : a \notin A_2\} \in \Theta.$$

(iv) If $\mu(A) = 0$ and $B \subseteq A$, then $B \in \Theta$.

Once we have defined the Lebesgue measure, we can introduce the concept of measurable function. We will say that the function $f : \mathbb{R} \to \mathbb{R}$ is Lebesgue measurable if the set $f^{-1}([a, b]) = \{x : a \leq f(x) \leq b\}$ is Lebesgue measurable for every open interval (a, b). Furthermore, from thinking about the sets with the Lebesgue measure equal to zero as negligible due to their insignificant size, we arrive at the idea of treating two measurable functions as equal as long as the set of points of the real line where they differ has zero measure. Indeed, when this happens, we say that both functions are the same almost everywhere. The same kind of technique is used to say that a function is less than or equal to another in almost every point.

If we now want to introduce the integral as an area, the idea is very simple: First, we define $\int_a^b \chi_A = m((a,b) \cap A)$, where A is supposed to be measurable and

$$\chi_A(t) = \begin{cases} 1 & t \in A \\ 0 & t \notin A \end{cases}$$

is the characteristic function associated with A. Next, we define

$$\int_a^b \sum_{k=0}^n \alpha_k \chi_{A_k} = \sum_{k=0}^n \alpha_k \int_a^b \chi_{A_k}.$$

And finally, for any bounded measurable function f, we define its upper integral $S_{[a,b]}(f) = \inf\left\{\int_a^b g : g = \sum_{k=1}^n a_k \chi_{I_k} \geq f\right\}$ and its lower integral $s_{[a,b]}(f) = -S_{[a,b]}(-f)$ and, when both values coincide, we say that f is Lebesgue integrable with integral

$$\int_a^b f = S_{[a,b]}(f) = s_{[a,b]}(f).$$

Of course, this definition can naturally be extended to consider improper integrals in the sense that both unbounded functions and infinite intervals of integration are admitted. It is not difficult to prove that every continuous function is measurable and integrable according to Lebesgue and, furthermore, its Lebesgue integral coincides with its Riemann integral, which also exists.

The Lebesgue integral has important properties related to some limit processes' interchanges. Concretely, the following results are especially important:

Theorem (Monotone convergence). Assume that $0 \leq f_1 \leq f_2 \leq \cdots$ are measurable functions defined on the measurable set E and that $f(x) = \lim_{n \to \infty} f_n(x)$ almost everywhere in E. Then, $\lim_{n \to \infty} \int_E f_n = \int_E f$.

Theorem (Dominated convergence). Assume that the functions f_i, $i = 1, 2, \ldots$, and g are measurable on the measurable set E and that $|f_i| \leq g$ almost everywhere in E, for $i = 1, 2, \ldots$. If $\int_E g < \infty$ and there is a function f such that $\lim_{i \to \infty} f_i(x) = f(x)$ almost everywhere in E, then $\lim_{i \to \infty} \int_E f_i = \int_E f$.

It is interesting to observe that a set is Lebesgue measurable if and only if it is the union of a Borel measurable set and a set of measure zero. This is why Borel stated that the only real contribution of H. Lebesgue (1875–1941) to the theory of measure was the precise definition of zero measure sets. Of course, this upset Lebesgue very much, who was also at the time financially burdened and overwhelmed by an excessive teaching load at a small French university. This type of dispute, whose background is to find out who did what and to what extent was it an original and exciting contribution, is very characteristic of mathematics. Rivalries worthy of a novel sometimes break out!

A Mathematical Model for Brownian Motion

In 1919, Professor I. A. Barnett (1894–1975), from Cincinnati University, commented to Wiener that an interesting field of study to which he could try to contribute was the generalization of the concept of probability to enlarge the class of spaces where this concept is applicable. Concretely, it would be interesting to apply it not only to spaces whose elementary events are points in the plane or the ordinary space but also to more complex cases such as, for example, that these were curves or functions. If we take into account that at that time the Theory of Probabilities had not been formulated in axiomatic terms and, therefore, the close relationship that exists between the concepts of probability, measure, and integral was not yet firmly established, it is clear that the task he had ahead was very complicated.

However, Wiener was ready to make a significant mathematical contribution. He had studied Lebesgue's integration theory in depth, first with Hardy and then alone, reading the books of F. W. Oswood (1864–1943), V. Volterra (1860–1940), M. Fréchet (1878–1973), and Lebesgue on topics of analysis. These books came into his hands thanks to his younger sister's relationship with G. M. Green, a promising mathematician at Harvard University who unfortunately died very young. His reading of Einstein and Smoluchowsky's work on Brownian motion and Perrin's book *Atoms* helped him develop a sensitivity to physics, which increased after discovering J. W. Gibbs's (1839–1903) statistical mechanics. Wiener admired Gibbs, with whom he shared the view that the statistical approach is an absolute necessity if we are to give an adequate description of macroscopic phenomena.

The problem proposed by Barnett already had a history of its own. Indeed, after the introduction of Lebesgue's theory of measure, there were several ways to generalize the concepts of measure and integral. One of them was introducing an abstract concept of measure that was independent of the dimension of the background space. In particular, Fréchet, founded on some earlier results by Radon (1887–1956) on an abstract concept of measure, had already raised the possibility of defining these concepts in spaces of arbitrary dimension. But no one had found concrete and non-trivial examples. Perhaps finding physical motivation could help.

Inspiration came to Wiener as he gazed at the Charles River from his MIT office. Watching the currents change, Wiener couldn't help but ramble on about the possible mathematical description of water flow. Would it be possible to find a suitable mathematical tool to account for the complex physical phenomena observed in the continuous movement of the waves, avoiding their excessive complexity? How could Newtonian mechanics, for example, describe the erratic motion of a particle carried along the turbulent waters of the river? Isn't one of the goals of science searching for order where (apparently) there is only disorder?

Fortunately, he had read that same year an article by the English mathematician P. J. Daniell (1889–1946) in which a very versatile formulation of the Lebesgue integral was given, a formulation that allowed the introduction of new concepts of integration. Wiener soon took advantage of Daniell's method and applied it to the study of certain general integrals.

Daniell's Integral

The main idea of Daniell's theory of integration is to assume that we know a concept of integral for a certain class of elementary functions and, from this, extend that integral to a much larger class of functions, so that the basic properties of the Lebesgue integral are maintained. A little more formally, things are as follows:

We admit that we have a set X and a vector space E of functions $f : X \to \mathbb{R}$, with the additional property that if the function f belongs to the given space, the function $|f|$ also belongs to E and that on this space we have defined a linear functional $I : E \to \mathbb{R}$ that satisfies the basic properties of the usual integral. Specifically, it must be verified that if $f \in E$ is such that $f \geq 0$, then $I(f) \geq 0$. Also, if $\{f_k\}_{k=1}^{\infty} \subseteq E$ is such that $f_1 \geq f_2 \geq \cdots \geq f_n \geq f_{n+1} \geq \cdots$ and for every $x \in X$ we have that $\lim_{n \to \infty} f_n(x) = 0$, then $\lim_{n \to \infty} I(f_n) = 0$.

Well, under these conditions, it is possible to extend the integral I to a class of functions that in principle is much larger than the elementary functions. To do this, the concept of zero measure set is first introduced. We say that the set $A \subseteq X$ has zero measure if for every $\varepsilon > 0$ there is a non-decreasing sequence of elementary functions $\{h_i\} \subseteq E$ verifying the inequalities: $h_i \geq 0$, $\sup_{i \in \mathbb{N}} h_i(x) \geq 1(x \in A)$ and $\sup_{i \in \mathbb{N}} I(h_i) \leq \varepsilon$. Finally, we will say that a property is satisfied almost everywhere in X whenever it is true for the points of the complement of a subset of X of zero measure. Let us now consider the class of functions L^+ that can be represented as the limit (at almost every point) of a non-decreasing sequence of elementary functions h_k that also has the additional property that the numerical sequence $I(h_k)$ is bounded. In these circumstances, given $f \in L^+$, we define its Daniell integral as $I(f) = \lim_{k \to \infty} I(h_k)$. At this point, it is evident that a proof of the fact that the value $I(f)$ does not depend on the chosen sequence h_k is necessary. Finally, the class L is defined as the set of functions that admit a representation of the type $f = f_1 - f_2$ with $f_1, f_2 \in L^+$. This is already a fairly general class of functions in comparison with the elementary functions from which we started. For the elements of L, the Daniell integral is defined by $\int_X f = I(f_1) - I(f_2)$.

Again, it is necessary to show that the definition of the integral does not depend on the chosen functions f_1, f_2. It is easy to check that L is a vector space. Furthermore, if we apply this construction starting from the usual characteristic functions, we obtain an integral equivalent to that of Lebesgue. We also obtain the Lebesgue integral if we take the continuous functions as elementary functions and the Riemann integral as their integral. Finally, if we use the Riemann–Stieltjes integral, we obtain the Lebesgue–Stieltjes integral.

Almost all of the important properties of the Lebesgue integral are inherited by the Daniell integral. These include the dominated convergence theorem, Fatou's lemma, the Riesz–Fischer theorem, and Fubini's theorem.

But Daniell's integral is much more suitable than Lebesgue's if we try to extend the concept to spaces of functions defined on infinite-dimensional domains. In particular, this theory is appropriate for defining integrals of linear functionals.

Wiener did something that a normal analyst of that time would never have attempted: He read an article by G. I. Taylor (1886–1975) on the physical phenomenon of turbulence. In this work, Taylor characterized turbulence in terms of certain means that depend on the complete path of each particle and Wiener tried to apply the integration process that he had introduced that same year to this phenomenon.

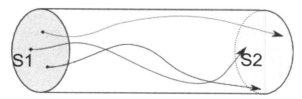

Let's imagine that we circulate a turbulent fluid through a pipe like the one in the figure and that, having fixed the sections S1 and S2 of the same (we suppose that the fluid moves from S1 towards S2), the section S1 is impregnated with a special dye with the property that each particle that passes through it is colored and leaves a trace from there. Furthermore, to distinguish the trajectories of the different particles, we will assume that each of them leaves a mark of a different color. If we want to study the phenomenon of turbulence, then we will have to measure certain properties (such as the average speed of a particle) that depend on each of the trajectories that appear in the fluid. Thus, we should know how to average the values of a certain functional (in our case, the average speed). This means that we must integrate a functional defined on the set of the trajectories of the particles that traverse S1, which is nothing more than a subset of the space of continuous functions C[0, 1].

Wiener was unsuccessful with the turbulence problem, but, as he was well informed about the Brownian motion, he decided to apply his ideas in this field and it was here that he obtained returns. Specifically, he provided a mathematical model of Brownian motion subjected to slight idealizations. In his model, Einstein had admitted the existence of a certain positive value τ such that the motion of the particle is, on consecutive time intervals of size τ, independent. From that point on, Wiener introduces the ideal hypothesis that Einstein's conditions are satisfied for all positive values τ. It is to this idealized movement that Wiener was able to apply the results obtained in his 1920 article on the Daniell integral.

Differential space. Let's see how Wiener constructs a measure on the space of continuous functions defined on the interval [0, 1]. The main idea is to use our knowledge of the probability that a particle that passes through the origin of coordinates at $t = 0$, at the instants $0 < t_1 < \cdots < t_n < 1$ is on the intervals $(\alpha_1, \beta_1), (\alpha_2, \beta_2), \ldots, (\alpha_n, \beta_n)$, respectively. This probability was, as we have commented previously, known. What Wiener observes is that the sets of continuous curves $x(t)$ that satisfy $x(0) = 0$ and $\alpha_i < x(t_i) < \beta_i$, $i = 1, 2, \ldots, n$ can be used to construct a topology on the space

$$C_0[0, 1] = \{x : [0, 1] \to \mathbb{R} : x(0) = 0, \text{ and } x(t) \text{ is continuous}\}$$

and that the probabilities assigned to these sets can therefore be used to define a Borel measure in this space. This measure is the so-called Wiener measure and it is with respect to it that Wiener first constructs an integral for certain elementary functions and, then, using Daniell's ideas, extends it to functionals.

To define the corresponding topology, Wiener selects a double sequence of points $\{t_{n,i}\}_{n=1,2,\ldots}^{i=1,2,\ldots,k_n} \subseteq [0,1]$ that verifies the following hypotheses:

(a) For each n, we have that $0 < t_{n,1} < t_{n,2} < \cdots < t_{n,k_n} < 1$.
(b) For each n, the inclusion $\{t_{n,i}\}_{i=1,2,\ldots,k_n} \subseteq \{t_{n+1,i}\}_{i=1,2,\ldots,k_{n+1}}$ is fulfilled.
(c) $\{t_{n,i}\}_{n=1,2,\ldots}^{i=1,2,\ldots,k_n}$ is a dense subset of $[0,1]$.

The next step is to introduce, for each $n \in \mathbb{N}$, the class π_n of subsets of $C_0[0,1]$ whose elements are the sets of the form

$$I = \{x \in C_0[0,1] : \alpha_i < x(t_{n,i}) < \beta_i, i = 1, 2, \ldots, k_n\}.$$

From condition (b) above it is clearly seen that inclusion $\pi_{n+1} \subseteq \pi_n$ occurs for every natural number n. Moreover, it is not difficult to check that the collection of subsets $\mathrm{B} = \bigcup_{n=1}^{\infty} \pi_n$ forms a basis for a topology on $C_0[0,1]$. The Wiener measure of the previous set is given by

$$W(I) = \int_{\alpha_1}^{\beta_1} \int_{\alpha_2}^{\beta_2} \cdots \int_{\alpha_n}^{\beta_n} p(0|x_1,t_1)p(x_1|x_2,t_2 - t_1) \times \cdots$$

$$\times p(x_{n-1}|x_n, t_n - t_{n-1})dx_1 dx_2 \cdots dx_n,$$

where

$$p(x|y,t) = \frac{1}{\sqrt{2\pi t}} \exp\left(-\frac{(y-x)^2}{2t}\right).$$

Obviously, this measure extends to the entire topology and, in fact, Wiener showed that it defines a Borel measure for $C_0[0,1]$ and that the measure of the whole space is 1, so that it is a measure that can be interpreted as a probability measure. Furthermore, Wiener shows that the set of possible Brownian motions (that is, the continuous mappings $x : [0,1] \rightarrow \mathbb{R}$ that are not differentiable at any point) has measure 1. With this, Wiener confirmed a phrase from Perrin, who had stated that

> Those who hear of curves without tangents or of functions without derivatives often think at the first instance that Nature presents no such complications nor even suggests them. The contrary, however, is true, and the logic of the mathematicians has kept them nearer to reality than the practical representations employed by physicists.

The space $C_0[0,1]$, endowed with the topology and the measure that we have just introduced, is called Wiener space or, as he himself baptized it, "differential space." On it, Wiener defined a concept of integral based on Daniell's ideas. To do this, he considers as a starting space the set E of simple functions defined from the elements of B, and as an integral of the simple function $f = \sum_{k=1}^{n} a_k \chi_{I_k}$, he takes $\int f dW = \sum_{k=1}^{n} a_k W(I_k)$.

In order to apply the Daniell integral extension process, Wiener had to show before that if $\{f_n\}_{n=1}^{\infty}$ is a sequence in E that converges to the function 0, then $\lim_{n \to \infty} \int f_n dW = 0$.

It is important to note that Wiener's space does not coincide, from the point of view of topology, with the space of continuous functions that vanish at the origin, $C_0[0,1]$, endowed with the usual topology since, for example, the convergence of a sequence of functions f_n to the function f occurs as soon as there is uniform convergence over the net of points $\{t_{n,k}\}_{n=0,1,\ldots}^{k=1,\ldots,k_n}$. This is a strictly weaker condition than uniform convergence over the entire interval $[0,1]$. To see it, it is enough to take f_n such that it vanishes on the points $\{t_{n,k}\}_{k=1,\ldots,k_n}$ and is equal to 1 on the midpoints $\left\{\frac{t_{n,k}+t_{n,k+1}}{2}\right\}_{k=1,\ldots,k_n-1}$. This sequence clearly converges to the zero function in Wiener space but does not converge anywhere in the usual space.

There are many ways to describe Brownian motion mathematically and, in fact, Wiener introduced several forms that he used in his later research. The underlying idea is the same in every case: It is about describing a stochastic process with some special characteristics. A stochastic process is a random variable whose values are not numbers but functions. But, what is a random variable? Suppose we design a random experiment whose sample space is the set Ω. That is, the elements of Ω are the elementary events of our experiment. For example, if the experiment consists of throwing a pair of dice into the air, the set Ω consists of the ordered pairs (a, b) with $a, b \in \{1, 2, 3, 4, 5, 6\}$. By its very nature, the random experiment determines on Ω a family of subsets M that are precisely those that represent a possible event or, in other words, those to which we can assign a probability. This, in analytical terms, means that the experiment is fully described by defining a function (the probability) $P : M \to [0, 1]$ that has the properties of a measure, with the additional condition that $P(\Omega) = 1$. If we define a function $X : \Omega \to \mathrm{R}$, which is measurable with respect to P, we will say that we have introduced a random variable. The random variable X will be analytically described by its probability distribution, which is the function F_X given by $F_X(t) = P(X \leq t) = P(\{a \in \Omega : X(a) \leq t\})$. In the case of a stochastic process, what we do is define an application that associates with each event $s \in \Omega$ a function $X(s)(t) = X(t, s)$ which we call the sample function of the stochastic process X. If the variable t runs through a numerical interval, we will say that the stochastic process is continuous and, on the contrary, if it can only take values on a discrete set — typically the set of integers or natural numbers — we will say that the process is discrete.

Before advancing in our explanation of stochastic processes, it is convenient to remember some elementary questions related to random variables, which are mathematical objects of a simpler nature. We have said that the distribution function $F_X(t) = P(X \leq t)$ completely determines the random variable X. What properties does it have? Obviously, $F_X(t)$ is an increasing function that, for $t \to -\infty$, tends to zero and, for $t \to \infty$, tends to one. Furthermore, any function $F : \mathbb{R} \to \mathbb{R}$ that verifies the above properties can be interpreted as the distribution function of a random variable. There are two types of random variables that appear with special frequency: continuous random variables, which are those whose distribution function can be expressed as $F_X(t) = \int_{-\infty}^{t} f_X(t)dt$ for a certain integrable function $f_X(t)$ (which we call the probability density function of the random variable), and the discrete random variables, which are those whose distribution function can be expressed as $F_X(t) = \sum_{t_j \leq t} P[X = t_j]$ for a certain sequence of real numbers $\{t_j\}_{j=0}^{\infty}$. So, when we fix the density function $f_X(t)$ or the probability function $P[X = t_j]$, we are fixing a type of random variable.

We have already commented that every random variable appears associated with a random experiment. The dynamic is as follows: The experience is carried out and then some type of measure associated with the result is taken. Then, it is worth asking, given that we know the distribution of a random variable X, if there is any way to find a numerical value μ that has the property that in a vast majority of cases, after conducting the experience and measuring its result, we will get a value that stays close to μ. Such a value exists. It is called mathematical expectation (or mean value) of the random variable X, it is denoted by $\mu = E(X)$, and its mathematical expression is $E(X) = \int_{-\infty}^{\infty} t f_X(t)dt$ if the random variable is continuous and $E(X) = \sum_{k=0}^{\infty} t_k P[X = t_k]$ if is discrete. On the other hand, the value of μ is not extremely significant on its own because it could happen that there are highly unbalanced areas in real terms. Due to this, for a better description of the behavior of the random variable X, it is also necessary to take into account its variance, which is given by $\sigma^2 = V(X) = E((X - \mu)^2)$.

Some examples of random variables. The more prominent examples of continuous and discrete random variables are the following:

(a) Discrete uniform: $P[X = k] = \frac{1}{n}; k = 1, \ldots, n$.

(b) Bernoulli: $P[X = 0] = p \in (0, 1)$ and $P[X = 1] = q = 1 - p$.

(c) Binomial: $P[X = k] = \dfrac{n!}{k!(n - k)!} p^k q^{n-k}$.

(d) Poisson: $P[X = k] = \dfrac{1}{k!}\lambda^k e^{-\lambda}$, where λ is a positive real number.

(e) Normal (or Gaussian): $P[X \le t] = \frac{1}{\sigma\sqrt{2\pi}} \int_{-\infty}^{t} e^{-\frac{1}{2}\left(\frac{s-\mu}{\sigma}\right)^2} ds$.

(f) Exponential: $P[X \le t] = \int_{0}^{t} \lambda e^{-\lambda s} ds$, where λ is a positive real number.

(g) Uniform: $P[X \le t] = \frac{1}{b-a} \int_{a}^{t} \chi_{[a,b]}(s)\, ds$, where $a < b$ are two real numbers.

Let us now consider the stochastic process $X(s)(t) = X(t, s)$. Associated with each instant t_0, we can introduce the random variable $X(t_0) : \Omega \to \mathbb{R}$ given by $X(t_0)(s) = X(t_0, s)$ and, henceforth, we can study its probability distribution.

Indeed, an equivalent way of introducing the stochastic process $X(t, s)$ is to specify the family of random variables $\{X(t)\}_{t \in T}$ and account for its main properties. For example, it is very reasonable, from the probabilistic point of view, that in order to study the stochastic process $\{X(t)\}_{t \in T}$, the joint distribution function of the random vector $(X(t_1), \ldots, X(t_n))$ is known for each choice of time values $t_1, t_2, \ldots, t_n$. This distribution is the function given by

$$F_{(X(t_1),\ldots,X(t_n))}(h_1, \ldots, h_n) = P(X(t_1) \le h_1, \ldots, X(t_n) \le h_n).$$

In particular, according to the point of view we have just exposed, every succession of random variables is an example of the stochastic process. Another way to introduce examples of stochastic processes is to consider sums of the type $X(t, s) = \sum_{k=0}^{n} A_k(s)\varphi_k(t)$, where the functions A_k are random variables of a certain type (for example, uniform, normal, or Poisson) and the φ_k are functions in the ordinary sense of the term. Even infinite sums of this type can be considered.

In the case of Brownian motion, the physical experiment consists, as we have seen, of randomly selecting a pollen particle (that we denote by s) at the instant of time $t = 0$ and then, for each instant $t > 0$, measuring the position of the particle at that precise moment, denoting the result by $X(t, s)$. However, this description does not provide us with any mathematical information about the process. If so, then where would Wiener's credit lie?

Wiener justified the need for his differential space by interpreting it in terms of a very specific stochastic process which we call the

Wiener process and whose interest lies precisely in the fact that the sample functions of that process are a very good idealization of the Brownian motion that is observed experimentally.

Specifically, Wiener showed that if $\{X_k\}_{k=0}^{\infty}$ is a sequence of independent random variables, all of them of Gaussian type with zero mean and variance 1, then the series

$$W(t) = \frac{X_0}{\sqrt{\pi}}t + \sum_{n=1}^{\infty}\left(\sum_{k=2^{n-1}}^{2^n-1} X_k\sqrt{\frac{2}{\pi}}\frac{\sin(kt)}{k}\right)$$

converges uniformly for $t \in [0, \pi)$, with probability 1, and thus defines a stochastic process $W(t)$. Furthermore, this process is such that

$$f_{(W(t_1),W(t_2),\ldots,W(t_n))}(h_1,\ldots,h_n)$$
$$= \frac{\exp\left(-\frac{1}{2}\left[\frac{h_1^2}{\alpha t_1} + \frac{(h_2-h_1)^2}{\alpha(t_2-t_1)} + \cdots + \frac{(h_n-h_{n-1})^2}{\alpha(t_n-t_{n-1})}\right]\right)}{\sqrt{(2\pi\alpha)^n t_1(t_2-t_1)\cdots(t_n-t_{n-1})}} \qquad (2.1)$$

for a certain positive constant α. Therefore, we can interpret its sample functions as the trajectories of the Brownian motion and, what's more, associated with this stochastic process, we inevitably find the measure of the differential space.

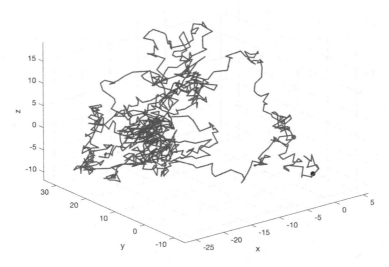

Example of Brownian motion in three-dimensional space.

Another idea, which was discovered later but is much more intuitive and also serves to mathematically define the Wiener process, is to start from the so-called simple random paths. These paths are a special type of discrete stochastic process. The starting point is to consider the sequence of random variables defined by $X(n) = h \cdot (D_1 + D_2 + \cdots + D_n)$, where $h > 0$ is a constant that we call the step of the process and $D_1, D_2, \ldots, D_n$ are independent random variables, which take the value $+1$ with probability $1/2$ and the value -1 also with probability $1/2$.

In more mundane terms, the variable $X(n)$ represents the position of a particle at the instant of time $n\tau$ when it is subjected to the following experiment: At the instant $n = 0$, the particle is located at the origin of the real line and changes its position $1/\tau$ times per second, with the condition that if the position of the particle at the instant $n\tau$ is given by $X(n)$, then $X(n + 1) = X(n) + 1$ with probability $1/2$ and $X(n + 1) = X(n) - 1$ also with probability $1/2$. From this sequence of random variables, we can define, for each value of the step h and of the unit of time τ, the continuous stochastic process

$$X(t) = h \cdot (D_1 + D_2 + \cdots + D_{[t/\tau]}) = hS_{[t/\tau]}.$$

If we now take $h = \sqrt{\alpha\tau}$ for a certain positive constant α and treat τ as a variable that we are going to make tend to zero, then we can define the stochastic process

$$W(t) = \lim_{\tau \to 0} X(t) = \lim_{\tau \to 0} \sqrt{\alpha\tau} \cdot S_{[t/\tau]} = \lim_{[t/\tau] \to \infty} \sqrt{\alpha t} \cdot \frac{S_{[t/\tau]}}{\sqrt{[t/\tau]}}.$$

Now, one of the most important theorems of probability theory, the Central Limit Theorem, states that if the random variables $X_1, X_2, \ldots, X_n, \ldots$ are independent and identically distributed, all with expectation $E(X_k) = \mu$ and variance $V(X_k) = \sigma^2 > 0$, then the sequence of random variables

$$Z_n = \frac{1}{\sigma\sqrt{n}}(S_n - n\mu) = \frac{1}{\sigma\sqrt{n}}(X_1 + X_2 + \cdots + X_n - n\mu)$$

satisfies that

$$\lim_{n \to \infty} P(Z_n \leq t) = \int_{-\infty}^{t} \frac{1}{\sqrt{2\pi}} e^{-\frac{s^2}{2}} \, ds.$$

If we apply this result to $W(t)$, we conclude that $W(t)$ follows a Gaussian (or normal) distribution with zero mean and variance equal to $\sqrt{\alpha t}$. Furthermore, due to the nature of simple random paths, it can easily be shown that the probability density function of the random vector $(W(t_1), W(t_2), \ldots, W(t_n))$ verifies formula (1), so that also in this case we can interpret the sample functions of the stochastic process $W(t)$ as the trajectories of Brownian motion.

Between 1920 and 1923, Wiener wrote several works about Brownian motion, with his article "Differential space," of 1923, being the most important one. However, it was not until 1930 that he gained a full understanding of the theory he had developed. This happened when he saw his ideas embedded into numerous physical phenomena. He included Brownian motion as part of a fundamental paper on generalized harmonic analysis, a work that we will discuss in the next chapter. Although these articles were excellent, the truth is that in the United States they received practically no attention. Fortunately, in Europe, some well-known mathematicians like Fréchet, and his student Levy in France, and Taylor and Hardy in England, understood the importance of his contributions. Wiener was aware of not being especially appreciated in his own country.

Furthermore, he was dependent on his father's opinions. The father, lacking his own criteria to assess the worth of his son's contributions, asked his Harvard mathematician colleagues, and they did not praise his son's work. Wiener suffered because the true value of his mathematical work was not appreciated even in the family home. Wiener would never recover from these impressions. He always thought that he was not cheered in his country. And he made it known to everyone, so much so that when, in 1966, two years after his death, the American Mathematical Society published a complete issue of the Bulletin dedicated to Wiener, one of his best students, N. Levinson, commented:

> Unfortunately Wiener did have grave doubts all of his professional life as to whether his colleagues, especially in the United States, valued his work, and, unwarranted as these doubts were, they were very real and disturbing to him.

The truth is that in the 1920s, American mathematicians tried to follow in the footsteps of G. D. Birkhoff (1884–1944) and O. Veblen

(1880–1960). Birkhoff was especially interested in classical analysis in H. Poincaré's (1854–1912) style, and Veblen was a brilliant topologist. Thus, although both were undoubtedly interested in physics, they concentrated all their attention on the then innovative efforts of the relativity and quantum mechanics theories and not on a subject as apparently isolated as the Brownian motion. This is why Wiener's work was not appreciated in the United States at that time. However, a moment of glory came with the publication in 1933 of the formalization of probability theory by A. N. Kolmogorov, since the international mathematical community became aware that Wiener had advanced in several essential respects beyond Kolmogorov. Furthermore, Wiener's integration techniques would later be successfully used in various disciplines, including quantum mechanics.

1920: The Strasbourg Congress and Banach Space Theory

In 1920, the Sixth International Congress of Mathematicians (ICM) would take place in Strasbourg, and Wiener wished to attend it. The 1916 ICM had been canceled due to the war, so eight years had already passed since the previous congress. Hence, this was the first opportunity for Wiener to attend an event of this importance. In reality, the 1920 Congress, which was organized by French mathematicians, was not going to have the character of the earlier ICMs because it excluded the participation of the former Central European powers (Germany, Austria, Hungary, and Bulgaria). This also motivated some renowned mathematicians from other countries to refuse to participate in the ICM.

In his autobiography, Wiener justifies his presence at the congress with the following words:

> The Germans were excluded as sort of punitive measure. In my mature, considered opinion, punitive measures are out of place in international scientific relations. Perhaps it would have been impossible to hold a truly international congress for another couple of years, but this delay would have been preferable to what actually did take place: the nationalization of a truly international institution. All that I can say for myself is that I was young (...) I was avid to seize the opportunity to revisit Europe with a certain small scientific status.

Wiener contacted Fréchet to propose that he accept him as a disciple during the months before the congress, which would take place in September. Fréchet not only accepted but also sent him an invitation letter and decided to spend part of his vacation working with him on common issues.

Wiener was attracted to the work Fréchet had done on topology and abstract spaces and saw in him a mathematician who would easily understand his ideas about measure and integration in spaces of functions. In fact, he used a good part of his new stay in Europe (first passing through England, then going to France and Belgium, and finally meeting with Fréchet) to develop his ideas on Brownian motion.

Once with Fréchet, Wiener thought that an interesting problem would be to characterize axiomatically an important class of abstract spaces in which Fréchet was working. This is how Wiener, entirely independently and almost simultaneously with Banach, introduced the concept of complete normed vector space (Banach space). In his autobiography, Wiener comments that at first Fréchet did not show much interest in his axiomatic research. Nevertheless,

> (...) A few weeks later, [Fréchet] became quite excited when he saw an article published by Stefan Banach in a Polish mathematical journal which contained results practically identical with those I had given, neither more nor less general. Banach's conception of his ideas and his publication of them were both a few months earlier than my own. There had, however, been no chance for communication between us, and the degree of originality of the two papers was identical. Thus, the two pieces of work, Banach's and my own, came for a time to be known as the theory of Banach-Wiener spaces.

In any case, Wiener abandoned the Banach space theory very quickly. Apparently, he did not have the character to work on issues under the extra pressure of competitiveness. In his own words:

> There were several motives which led me to abandon this brain child, of whom I was at least one of the parents. The first was that I did not like to be hurried or to watch the literature day by day to be sure that neither Banach nor any of his Polish followers had published some important result before me. All mathematical work is done under sufficient pressure, and its increase by such a fortuitous competitive element is intolerable to me.

In addition, Wiener also justified his decision based on aesthetic issues. Indeed, on numerous occasions (and in a very particular way, in his biographical writings), he argued in favor of a vision of mathematical activity as something full of aesthetic motivations, comparable to any of the so-called "fine arts"; according to him, every mathematician has, as happens with artists, his own muses. Thus, for him, an interesting mathematical theory must have some connection to physical reality or applications. However, in its early days, the Banach space theory lacked these motivations. Moreover, he believed that this theory was initially the pasture of numerous mathematicians eager to publish with ease, interested more than anything else in generalization after generalization. This was also annoying to him. Finally, his "differential space," although very close to Banach's spaces, was better motivated from the physical point of view, and it was his own creation, which he had no reason to share. That was also a rising value for Wiener. For this reason, he still devoted quite a bit of effort in the direction of better understanding Brownian motion and similar phenomena.

In his autobiography, Wiener thanks Fréchet who, although he initially did not show great interest in his research, did something important since he put him in contact with one of his best students, P. Levy (1886–1971). He was probably the only mathematician who understood from the beginning the full extent of the results that Wiener had in his hands at that time. Levy had studied the works of R. E. Gateaux (1889–1914) on functions of an infinite number of variables and on integration in infinite-dimensional spaces. Wiener himself admits that it took him some trouble to convince Levy that his ideas were going in another direction and were more suitable. But when he did, Levy became his ally and his defender for the rest of his life.

As we have already mentioned, the Strasbourg Congress had gone down in the history of the ICM as the one with the fewest attendees in the 20th century. However, some figures of French, English, and American mathematics gathered there. Among the Americans, Wiener remembers L. P. Eisenhart (1876–1965) of Princeton, L. E. Dickson (1874–1954) of Chicago, and S. Lefschetz (1884–1972) of Kansas. From France, apart from Fréchet, he remembers Jordan (who was then in his 90s, despite which he accompanied guests on their country walks) and J. Hadamard (1865–1963). From England, he mentions Sir A. G. Greenhill (1847–1927).

At the end of the congress, Wiener found that maritime transport with America was blocked as a result of a breakdown that his ship had suffered and so he had to wait the entire month of September and part of October until the connection was reestablished. Consequently, he came to MIT with the course year already started, a little worried about what the university authorities might be thinking of him. However, both this and many other trips that he would make throughout his life left a good taste in his mouth.

> I came back from Europe with renewed and enlarged inspiration. For all of the defects of my French, I had lived in France and had for the first time established a contact with my French colleagues. Both in France and in England I found that I occupied a more important position than at home, and I had work pending which seemed (at least to myself, if to few others) the promising beginning of a career in mathematics.

Indeed, as soon as he arrived at MIT, he began to write his papers on the Brownian motion and, advised by Professor E. B. Wilson (1879–1964), sent two papers to the Proceedings of the American National Academy of Sciences, where they were published in 1921.

Potential Theory: The First Act of Rebellion

There are many kinds of scientists. In physics, one way to easily distinguish between one and the other is to find out whether they are applied or theoretical physicists. It is even easy to differentiate between these types of physicists simply by observing their schedules and habits. On this point, it is instructive to read the amusing comments of the Nobel Laureate in physics L. Lederman (1922–), an "experimenter," in his book *The God Particle*, where he affirms, referring exclusively to the world of particle physicists, that:

> Theorists tend to come in late to work, attend grueling symposiums on Greek islands or Swiss mountaintops, take real vacations, and are at home to take out the garbage much more frequently. They tend to worry about insomnia (...) Experimenters don't come in late — they never went home. During an intense period of lab work, the outside world vanishes and the obsession is total. Sleep is when you can curl up on the accelerator floor for an hour.

However, in mathematics, the fundamental distinction may lie elsewhere. There are indeed theoretical and applied mathematicians, but that is not essential. Some mathematicians work on developing a theory or on the unification of one or more subjects, and on the other side are those who are dedicated to solving specific problems. It is also important to know if a mathematician always works on a single topic to which he dedicates exclusive attention or if, on the contrary, he needs to think about several things simultaneously and go from one place to another, recreating himself, clearing himself of a problem while thinking about another. Wiener was of this latter type of mathematician. Moreover, he was interested in problems that had some physical or engineering motivation. Thus, during the 1920s, he worked on several subjects at the same time.

We have already commented on two of them: the Brownian motion and the Banach space theory. Other topics were harmonic analysis and potential theory. We will discuss harmonic analysis in detail in the next chapter, so now we will explain his contribution to potential theory. Wiener only worked on this subject for a short time, from 1923 to 1925, but his ideas had a considerable impact.

The question he faced was the solution to the Dirichlet problem for domains whose border is not smooth. Dirichlet's classical problem consists of searching for solutions to the equation

$$\Delta u = \frac{\partial^2 u}{\partial^2 x_1} + \frac{\partial^2 u}{\partial^2 x_2} + \frac{\partial^2 u}{\partial^2 x_3} = 0 \text{ for } (x_1, x_2, x_3) \in \Omega \text{ and}$$

$$u(x_1, x_2, x_3) = f(x_1, x_2, x_3) \text{ for } (x_1, x_2, x_3) \in \partial\Omega,$$

where Ω is a domain (that is, an open and connected subset) of the three-dimensional Euclidean space $\mathbb{R}^3$ and $f : \partial\Omega \to \mathbb{R}$ is a continuous function on the boundary of Ω.

Of course, we look for solutions u that are twice differentiable on the interior points of Ω and are continuous on the closure of Ω. It is also clear that the problem has analogous formulations for any finite number n of variables. In fact, Wiener contributed with new ideas, especially for the case $n \geq 3$. The case $n = 2$ is different because it is susceptible to being studied with complex variable techniques and it was the first one for which convincing and definitive answers were given.

The equation $\Delta u = 0$, which is called Laplace's equation, appears in many physical situations. Furthermore, its history is very much a model of the history of the relationship between physics and mathematics, especially between physics and analysis. The best-known example of a physical problem described by this equation is the stationary distribution of temperatures in a region of space Ω on whose boundary the temperature value has been set at each point and whose interior lacks heat sources. Thanks to Fourier's work, we know that the identity that describes the heat flow is in this context

$$\frac{\partial u}{\partial t} = K \left(\frac{\partial^2 u}{\partial^2 x_1} + \frac{\partial^2 u}{\partial^2 x_2} + \frac{\partial^2 u}{\partial^2 x_3} \right).$$

Hence, if we let the system stabilize, the equation that describes the stationary temperature distribution results from imposing $\frac{\partial u}{\partial t} = 0$, which leads to Laplace equation in three-dimensional space,

$$\Delta u = \frac{\partial^2 u}{\partial^2 x_1} + \frac{\partial^2 u}{\partial^2 x_2} + \frac{\partial^2 u}{\partial^2 x_3} = 0.$$

The solutions of this equation are called harmonic functions and potential theory is the branch of mathematical analysis (and mathematical physics) that studies them. Thanks to a fundamental result of this theory, the Maximum Principle, it can be proved very easily that if there is any solution to the Dirichlet problem, it is unique. That is, only one solution to the problem can exist. On the other hand, it was discovered that certain domains Ω exist for which the problem lacks solutions in the sense that we have given because the only candidate for a solution does not satisfy $u(x_1, x_2, x_3) = f(x_1, x_2, x_3)$ everywhere in $\partial\Omega$. As a consequence, for each function f defined on the boundary of Ω, the possibility of finding a harmonic function $u = u_f$ that verifies $\lim_{p \in \Omega, p \to p_0} u(p) = f(p_0)$ for the largest possible set of points $p_0 \in \partial\Omega$ was raised. The problem, formulated in this way, is the so-called generalized Dirichlet problem.

Wiener's interest in potential theory, specifically in this general problem of Dirichlet, grew out of his conversations with O. D. Kellogg (1878–1932) in the early 1920s. Kellogg, who had started in potential theory at the very beginning of the 20th century under the command of Hilbert, who in turn had solved Dirichlet's classical problem in an absolutely spectacular way in 1900, was very happy when Wiener expressed his interest in potential theory, and he proposed to Wiener one of the most important open problems of the theory at that time.

The aim was to characterize the regular points of the domain: The points $p_0 \in \partial\Omega$ for which $\lim_{p\in\Omega, p\to p_0} u(p) = f(p_0)$, where u represents the solution of the generalized Dirichlet problem associated with the function f.

S. Zaremba (1863–1942), a Polish mathematician, had obtained, on the one hand, certain sufficient conditions and, on the other hand, necessary conditions for the regularity of a point. But he had not found any condition that was simultaneously necessary and sufficient. The problem was considered very difficult, so it is very likely that Kellogg was not hopeful about Wiener's success — who after all was not an expert in the subject. However, the potential theory was one of those subjects that could attract Wiener's imagination and keep him captivated for a long time since it was a theory with important physical content. That was precisely what happened and perhaps often what happened to him on other occasions: that the physical motivation of a complex mathematical problem was the fulcrum and the key to his inspiration when approaching imaginative solutions, bringing the work to fruition.

Zaremba's problem

Wiener not only solves the generalized Dirichlet problem by uniquely associating with each function f defined in $\partial\Omega$ a function $u = u_f$ that satisfies Laplace's equation $\Delta u = 0$ in Ω but also provides a geometric characterization of the regular points $p_0 \in \partial\Omega$. The solution u_f is obtained from f as follows: First, he introduces the concepts of superharmonic function — a function w such that $-\Delta w \geq 0$ on Ω — and subharmonic — a function w such that $-\Delta w \leq 0$ on Ω — and from them, he defines the functions

$$\overline{H}_f(x) = \inf\{w(x) : w \text{ is superharmonic in } \Omega \text{ and } w \geq f \text{ in } \partial\Omega\},$$

$$\underline{H}_f(x) = \sup\{w(x) : w \text{ is subharmonic in } \Omega \text{ and } w \leq f \text{ in } \partial\Omega\}.$$

He then proves that if f is continuous in $\partial\Omega$, then $\overline{H}_f$ and $\underline{H}_f$ are harmonic functions and coincide on Ω. Thus, he defines $u_f = \overline{H}_f$ in Ω and studies the points of $\partial\Omega$ for which $\lim_{p\in\Omega, p\to p_0} u_f(p) = f(p_0)$ (the regular points).

To characterize the regular points, Wiener introduces an idea that will be of vital importance to the theory of potential: the concept of capacity. If E is a bounded Borel set, its capacity is given by

$$C(E) = \inf \left\{ \int_{\mathbf{R}^n} \|\nabla \phi\|^2 : \phi \in C_0^\infty(\mathbf{R}^n), \phi_{|E} \geq 1 \right\}.$$

Finally, Wiener shows that the point $x_0 \in \partial \Omega$ is regular if and only if

$$\int_0^1 \frac{C((\mathbf{R}^n \backslash \Omega) \cap B(x_0, r))\, dr}{r^{n-2}} \frac{dr}{r} = +\infty.$$

In particular, if it is locally Lipschitz, all its points are regular and therefore for this type of domain the classical Dirichlet problem is solvable.

When Wiener showed Kellogg his work, a conflict ensued. Apparently, the results were "too" good. The specific problem was that two Princeton mathematicians, friends of Kellogg, were about to defend their doctoral theses, and their work was "displaced" by Wiener's. So, Kellogg asked him to delay publishing his results. Wiener was offended. He expected congratulations and what he received was more or less a "forget what you have done and let others step on your work because this is not your subject." Obviously, Wiener was unwilling to give in, so he sent his article as quickly as possible, and he sent it to a journal in which it would be the safest to publish: the new journal that his department at MIT had founded and for which he had been made the first director. In addition, he was quick to write a couple of articles more on potential theory, one of them with his MIT colleague Professor H. B. Phillips, with whom he developed a method of numerical resolution based on the use of finite differences that would later inspire him to propose the creation of digital computers. The work in which he would finally give the definitive solution to the Zaremba problem was also published under the pressure of competitiveness, as he realized that both Lebesgue and a student of his, named G. Bouligand, were sending numerous papers on this issue to the *Comptes Rendus* — an important French journal of which Lebesgue was editor — and that most likely one of them would soon come to the solution. Indeed, Lebesgue simultaneously received Wiener's and Bouligand's paper, both with the correct solution. The two works appeared in the same volume of the journal, in 1924, with an introductory comment by Lebesgue.

Although some biographers of Wiener claim that he overreacted to Kellogg's demands, the truth is that he had justification. He not only considered the request unethical but was also aware of the importance of his theorem and that it would be a fundamental boost to his prestige as a mathematician. In addition, Wiener did not see himself at that time (and he was not) as being supported by the mathematicians of his country — which was the case with any of Kellogg's protégés — and he was beginning to feel a real need to strengthen his position at MIT, since just at that time he was thinking of getting married.

It is in this way that Wiener went from being a child prodigy who had read a thesis in philosophy of mathematics and who was expected to one day make contributions of interest in philosophy to being a mathematician with a certain prestige, who had already done something important. So began his long career as a mathematician.

Chapter 3

Life Among Engineers

MIT is Transformed

When Wiener arrived, MIT was an institution whose main objective was to train engineers who would later perform competent work in the labor market. That is why the mathematics department was not conceived to carry out research, but, on the contrary, it was a department dedicated almost exclusively to teaching. The teachers were overloaded with classes and, in the words of S. J. Heims, "research in the areas of pure mathematics and fundamental physics was an anomaly." Obviously, this situation could not be to Wiener's taste since he wanted with all his might to make some critical contribution to the advancement of mathematics. He was not going to resign himself to teaching his classes and nothing else.

Fortunately, there were more professors who considered the creation of strong departments in the areas of mathematics and fundamental physics to be of paramount importance to the institution. Moreover, the historical situation was conducive to change because precisely from the mid-1920s, those responsible for the management of American scientific policy felt very attracted by the then newborn quantum physics, which obviously had a military interest. Indeed, they realized that Einstein's equation establishes a total equivalence between mass and energy. Furthermore, the equation revealed that a relatively small amount of mass is equivalent to an exorbitant energy amount. Therefore, they faced a fundamental problem of theoretical physics: How to transform mass into energy? The solution to this problem passed, no doubt, through a systematic and deep study of the fundamental nature of matter, and that is quantum physics. They could not prevent Europe from being at the forefront of theoretical physics advances. However, they were forced to make a significant effort to get the United States on the theoretical research bandwagon. The best possible method was the incorporation of quantum physics into American universities and the research institutes of some of the most prestigious scientists in Europe, who could later "create a school." Thus, something similar to what is currently done with elite athletes occurred: There were signings. With the support of the fortune of the Rockefeller and Guggenheim foundations and the General Education Board, several million dollars were invested in scholarship programs and research stays with which, among other things, mathematicians like Wiener traveled to Göttingen (1926) and physicists like Max Born visited MIT (1925).

Indeed, the first serious attempt to strengthen the departments of mathematics and fundamental physics at MIT took place in 1925, when a group of professors tried to attract Max Born, intending to have him found a department of theoretical physics at MIT. Coincidentally, during that time, Born and Wiener worked together on a paper on the mathematical foundations of quantum mechanics, but Born ultimately did not stick around.

Karl T. Compton
(1887–1954).
Courtesy: Public domain,
Wikimedia Commons.

The real change would take place after 1930 when Karl T. Compton (1887–1954) assumed the institution's presidency. Compton came from Princeton University, where he had defended his doctoral thesis in 1912 and had carried out an enormous amount of research work in particle physics, becoming director of the physics department, a position from which he had managed to attract some of the most prominent European minds, such as the mathematicians J. von Neumann and H. Weyl.

Compton came to MIT right in the middle of the great economic crisis that had caused the collapse of the US stock market in 1929. Under such circumstances, society's appreciation for basic science research was rocky, so one of MIT's main objectives was to recover lost prestige. That meant, among other things, carrying out a major educational reform that would significantly affect all engineering careers. In addition, Compton thought it necessary for professors dedicated to basic research to see their salaries increased to the level of those received by their peers at other prestigious institutions, such as Princeton or Harvard. MIT, however, could not make it happen in a few years, precisely because of the economic situation that the country was in. However, later it would fulfill its promise, and those who had faith in it and "held out at MIT" during the hard times (including Wiener) were not disappointed.

Wiener Gets Married

Margaret and Norbert Wiener with their two daughters. Courtesy: MIT Museum.

Due to the iron discipline with which he had been brought up from early childhood, Wiener had an insecure character. He also lacked the necessary social skills to define himself as an independent person. How could he establish a relationship for himself? To make matters worse, he also had no freedom in his relationships with the opposite gender since every time he brought a friend into the house, she was subjected to an exhaustive analysis by Wiener's parents (and sisters), from which she generally came out wrong. According to Wiener, they "exercised a complete veto right" and

> This veto was governed more by what my parents conceived to be the girls' reaction to the rest of the family than by any factor directly concerning me,

which only made things worse.

Wiener's parents got into the habit of throwing parties at home. They were mostly attended by his father's students and colleagues and some visiting professors when they were at Harvard. These parties were generally held on Sundays. They drank tea and ate a light meal. Wiener's and his sisters' participation in Sunday tea was supposed to serve as a "training ground" in which they would improve their social skills. Indeed, it was there that Wiener met Marguerite Engemann (1894–1989), who would later become his wife.

Marguerite was the younger sister of one of Norbert's students at MIT. In addition, she had studied Romance languages at Harvard and, in particular, had been a student of Leo Wiener in his Russian literature classes. Marguerite was of German origin. She had immigrated to the United States from Silesia (then in Prussia and now in Poland) when she was fourteen years old. Her mother, Hedwig Engemann, who belonged to a farming family in the Alps, had married an innkeeper who, after economic success, suddenly died when she was not yet forty years old. Her husband's death is what motivated her to immigrate to the United States and, from there, bring her children one by one. She did not succeed with everyone. The eldest, one of the sisters, stayed in Germany in the care of relatives. The rest — Marguerite and three brothers — were with their mother. Marguerite (whom everyone called Margaret except Wiener's parents) had received a very German-style upbringing. She had the Victorian morals and refined ways that Wiener's parents liked so much. Physically, she did not look Germanic but rather Latin. She had dark hair (almost black), olive skin, and hazel eyes. She was two months older than Wiener, a little shorter, and had a strong temper. Leo and Bertha saw in her the ideal companion for the young Wiener, whom they considered "socially inept" and, therefore, in need of a wife with a strong character and who knew how to "direct" his social life.

However, Wiener had other ideas in mind. Shortly after entering MIT, he had met a beautiful French student whom he had courted for a whole year until she told him that she was engaged to another man. Wiener took it pretty badly.

Wiener did not like that his parents showed such a clear interest in Marguerite and this, naturally, had the opposite effect to that desired by them:

> I felt so embarrassed at their obvious reaction to Margaret's favor that my response was to stay away from her for a while.

Wiener would spend the summers of 1922, 1923, 1924, and 1925 in Europe, making scientific contacts. During that time, he still had the opportunity to fall in love with another woman: the astrophysicist

Cecilia Payne-Gaposchkin (1900–1979). But she was not interested in love affairs since her greatest desire was to develop a brilliant scientific career, and that seemed incompatible with marriage.

Particularly successful (and we will write about this later) was his visit to Göttingen in the summer of 1925. There, he would give a lecture at a mathematical seminar directed by Hilbert, which, in his opinion, had a significant impact.

After this visit, he was invited by Born, Hilbert, and Courant (the new head of the mathematics department) to spend the academic year of 1926 there. To do this, Wiener had to get a scholarship in the United States and permission from the MIT authorities. Once the newly created Guggenheim Foundation awarded a grant for his stay, there were no problems from MIT. He was excited about this trip. For some reason, he had deduced from his conversations with Courant that his stay in Göttingen was going to be important not only for him but also to Göttingen's people and that he would be received with "honors."

Wiener was very happy. His departure as a Guggenheim fellow gave him so much security and hope that he was finally going to be recognized (if not in the United States, at least in Europe) as a first-rate mathematician that he did not hesitate to ask Margaret to marry him. She accepted. They planned to marry in the spring of 1926. The wedding was held in March in Philadelphia, and then they had a brief honeymoon in Atlantic City. Margaret had to go back to her classes in Pennsylvania (where she had acquired a position as a language teacher at a small university), so she accompanied Wiener only to the embarkation for Europe. They agreed to meet again in Germany at the end of June.

However, Wiener's dream would quickly turn into a nightmare. To begin with, on the occasion of the publication by the American press of a list of Guggenheim fellows who would travel to Europe that year, Wiener made the mistake of granting an interview, which would later be translated and published in Germany, in which he explained the reasons for his trip. In that interview, as he later admitted, "he was too talkative," giving himself greater importance than could honestly be attributed, and this had the unexpected and unwanted effect of the German authorities noticing him and tying up loose ends. Indeed, it was precisely then they realized that the Wiener who came to Germany to benefit from the good state of mathematics there was none other than the son of Leo Wiener, a prestigious professor of Slavic languages living in the United States who had repeatedly expressed his strongest rejection of the actions of Germany during the First World War. How was it possible that Germany welcomed this professor and allowed him to give lectures at such a prestigious university as Göttingen? Obviously, these thoughts have to be placed in the context of the German nationalism that was in full swing at that time and, therefore, these questions touched the pride of many people there.

Quickly, the Kurator (that is, the government representative at the university) contacted Courant and asked him to keep Wiener's presence as low key as possible. Courant saw Wiener as a very talented mathematician but did not consider the statements he had made to the press to be justified. Moreover, he was in a particularly delicate situation because his main objective was, at that time, to obtain a significant amount of money from the different administrations that would allow him to found a Mathematical Institute in Göttingen, which was soon to become a world reference center, so he couldn't argue too much to defend Wiener.

Thus, when Wiener arrived in Göttingen, he found a very cold Courant, one would almost say distant, who did not offer him the official welcome that he believed he deserved and who only fulfilled a minimum part of what was promised only a few months ago, reluctantly granting him only a few privileges (among others, the assignment of an assistant to prepare his lectures). In the case of Wiener — who had a very touchy and, in some respects, weak personality — this had the effect that he could not satisfactorily prepare his lectures — which, on the other hand, were hardly attended by any. The classes were badly received from a technical point of view (that is, the quality of the material presented), linguistically (he addressed them in terrible German), and, of course, by the public that attended them.

This episode was collected in a biography of Courant that, many years later, would be published by C. Reid (1918–2010) and which was based on direct interviews with the protagonist and numerous people around him.

> Courant was disposed, he told me, "to do something for Wiener" because the American was a cousin of Leon Lichtenstein, a good friend [of his]. But it was only with great difficulty

that he was able to persuade about twenty students to attend
Wiener's lectures. As time went on, that number dwindled so
embarrassingly that at one point, he had to pay a student [out
of his own pocket] to attend.

Obviously, Wiener never knew anything about this. On the contrary, he felt mistreated and deceived by Courant, whom he accused
of having used him for political purposes. Specifically, he was convinced that Courant had made his visit coincide with that of the
most famous American mathematician of the time, G. D. Birkhoff
(1884–1944), from whom he expected unconditional support for the
Göttingen Mathematical Institute, in the hope that this would posi-
tively influence Birkhoff. What Courant could not know (and Wiener
quickly pointed this out to him) is that his relationship with Birkhoff
was uneasy. Wiener believed that Birkhoff resented him because he
saw in him an equal and, possibly, a tough competitor in the future.
These obsessions were typical of him. Indeed, Wiener was quite capable of convincing himself of anything even if the facts showed otherwise. It seems that Birkhoff had no problem with Wiener (or, at least,
he is not known to have ever performed any maneuvers against him).
On the other hand, this would not be the only insult received from
Courant in Wiener's eyes. In 1928, Courant published a joint article with K. O. Friedrichs (1901–1982) and H. Lewy (1904–1988) in
Mathematische Annalen in which, among other things, he presented a
probabilistic interpretation of the equations in finite differences that
was originally due Wiener and Phillips; but they were not mentioned,
for which Wiener protested vigorously. Apparently, this attributing
to oneself the ideas of others was something relatively common at
that time in Göttingen. They were very careless with the work of
others because they considered themselves the center of the mathematical world. Not only did Courant do it but many others did as
well, such as Hilbert. They called it "nostrification" and, according
to Reid, there were three levels: conscious nostrification, unconscious
nostrification — which, according to Friedrichs, seemed to be the case
with Courant and Wiener, since, after the latter's protests, Courant
was always cautious to mention him — and self-nostrification. The
latter case consists of the publication of an extremely important idea
that, with time, one discovers had already appeared in a previous
work of oneself.

R. Courant (1888–1972) and G. D. Birkhoff (1884–1944).
Courtesy: Public domain, Wikimedia Commons.

When Margaret appeared in Göttingen, she found Wiener desolate, depressed, and isolated. Her reaction, naturally, was to comfort him and help him reestablish some of the relationships he had there. To make matters worse, Wiener's parents had planned to pay a visit to the newlyweds. Presumably, given the type of relationship they had always had with Wiener, one of their motivations was to exercise some control on the couple. They also wanted to share in their son's success and see him enjoy his sweet moment, just married, and finally a recognized person in his profession. The idea of the visit haunted Wiener. How could he explain to his father that his trip was proving a total failure and that the responsibility was largely his? But he had to. His parents' visit was therefore not pleasant. Instead, it was a nightmarish week in which the father was more concerned about the negative opinion they had of Leo in Germany than about the problems his son was going through. The couple finally decided to travel to Switzerland for the holidays and then escape to Italy for a few weeks to relax and enjoy a real honeymoon. They then attended a congress organized by the German Association for the Advancement of Science in Düsseldorf. From there, they traveled to Copenhagen, where Harald Bohr (1887–1951) awaited them, with whom Wiener would work during the last months of his stay in Europe as a Guggenheim fellow.

As soon as they returned to the United States, the Wieners looked for a home, and shortly (in February 1927) their first daughter, Barbara, was born. In December 1929, they had another girl, whom they named Margaret. They had no more children. Regarding his qualities as a father, S. J. Heims would comment the following:

Norbert Wiener took his parental responsibilities seriously and, in particular, tried to avoid the model of education that his father had imposed on him. To assess his parenting behavior, one must weigh his alternative exuberance and depression and his poorly disciplined emotional life, with his lively and imaginative company and warm concern for well-being of his daughters.

Life at MIT

Wiener found a warm home to take refuge in at MIT. At first, he lived with his parents, and he used the summers to travel to Europe, where he had good contacts. Then, he would live with his wife (and his daughters, when they arrived) and broaden the spectrum of his travels, including places as diverse as China, India, and Mexico. A university professor's normal life usually allows these trips either by attending conferences or as a visiting professor. Wiener thought that if he could work at home, he could work anywhere else as well, and he loved to travel. He enjoyed languages and contact with cultures different from his own. In fact, in his description of Wiener, H. Freudenthal (1905–1990) emphasizes that he "knew many languages but it was very difficult to understand him in any of them."

Over time, between Wiener and MIT, there was a rare symbiosis according to which the institution would always extend kind and understanding treatment towards Wiener — who, as a professor and as a researcher, was most peculiar — and, in return, it would receive much of its prestige from Wiener and his excellent research results. This exchange was beneficial to MIT, especially during the great change in the 1930s, when the institution went from being a center for "training engineers" to be a center for research (in all areas of science) of the first order. Still, it took Wiener some time to land a professorship at MIT. This was due to two equally significant factors: The first important results of Wiener, on Brownian motion, took at least a decade to be assimilated in the United States and, on the other hand, the 1930s were times of economic crisis, and consequently, academic careers were significantly slowed down during that decade. Even so, Wiener only tried once, shortly after marrying and having his first daughter, to change universities, applying to two professorships abroad (one at King's College in London and the other one at an Australian university), but he failed to win them.

Wiener had some fixations. He loved walking around MIT while he thought about his problems. Usually, he would approach whomever he met on his walks — be it a colleague, a student, or a janitor, that did not matter to him — and start a conversation about some of the issues he was working on. If he did not see anyone, he would walk into an acquaintance's office and deliver his monologue. Dirk Struik (1894–2000) said that "sometimes he would tell you something completely nonsensical, incomprehensible, but other times it was absolutely luminous, clear, intelligent ... almost prophetic." Ultimately, Wiener needed to communicate his ideas and did not hesitate to do so with anyone around. Sometime later, he would have problems because part of his investigations relating to military matters would be subject to state secrecy for a period of time and, therefore, could not be communicated. That was something he did not handle well. Wiener required the walks to think about mathematics, especially when he got stuck on some idea or demonstration. If he needed to write something in the middle of a mathematical conversation, he did not hesitate to enter the first office he saw (many times without calling) and, usually to the astonishment of the professor who occupied that room, start writing whatever it was on the blackboard. To compound the situation, he was shortsighted and wildly clueless. It was common for him to request several times to introduce himself to the same person. On some occasions, he walked into the wrong class, and, unaware of it, he gave his lecture without flinching. On other occasions, when it was time for his class, his students would go out to look for him in the corridors and the MIT offices, interrupt his conversation with a teacher, and take him to the classroom to give his lesson. His relationship with the students was good. If they asked him to give a special lecture, he always accepted and, in fact, gave it willingly.

His ability to concentrate was extremely high. He often believed that he had shown a result on the blackboard without actually writing anything. Once, a student delicately asked him to retest the result he had just explained, and Wiener was delighted to reply that there was no problem with it. He went to the board, concentrated for a while, and (again not writing anything at all) said, "That's it."

Now, none of the above means that he was a poor teacher. He was somewhat irregular. Similarly, he could bore his audience incredibly (doing many calculations that nobody could understand) or keep a whole class in suspense during the whole lesson, pending what he was going to do or say, intrigued and surprised by the arguments. The students loved him. He showed interest in his students, and, in connection with this, several significant anecdotes can be told. For example, N. Levinson commented that while attending a Wiener graduate seminar in 1933, the following happened to him:

Norman Levinson
(1912–1975).
Public domain,
McTutor.

He would actually carry on his research at the blackboard. As soon as I displayed a slight comprehension of what he was doing, he handed me the manuscript of Paley-Wiener for revision. I found a gap in a proof and proved a lemma to set it right. Wiener thereupon sat down at his typewriter, typed my lemma, affixed my name and sent it off to a journal. A prominent professor does not often act as secretary for a young student. He convinced me to change my course from electrical engineering to mathematics. He then went to visit my parents, unschooled immigrant working people living in a run-down ghetto community, to assure them about my future in mathematics. He came to see them a number of times during the next five years to reassure them until he finally found a permanent position for me. (In those depression years positions were very scarce).

Levinson became a highly respected mathematician and, among other things, was chairman of the mathematics department at MIT for a time and received the Bôcher Prize from the American Mathematical Society in 1954.

Not everyone was included in Wiener's round of visits (which his colleagues called the "Wienerweg" because, in German, the word "weg" means "path"). Indeed, some were proud that Wiener visited them. An example of this was Julius Stratton (1901–1994), who began receiving visits from Wiener while he was still a student of his and continued to receive them even when he was president of MIT, even then enjoying the interruptions:

> Already in the president's office I still liked very much that Wiener entered as before ... Of course, it is customary that when you get to the president's anteroom you wonder if he is busy. Norbert did not. He simply entered, whoever was there, interrupted and began to speak. It wasn't bad manners. He was simply self-absorbed.

Another MIT president who would welcome Wiener's visits was Jerome Wiesner, who would acknowledge that,

> Whatever was on his mind, for me and many others Norbert Wiener's visit was one of the highlights of the day at MIT. For those of us working in the solitary isolation of Building 20, Norbert's visits were especially welcome because they represented one of our best links to the main building.

Fourier Analysis

We have already explained that Wiener showed great interest in the Lebesgue integral. Furthermore, in the historical introduction that we made to this theory in the previous chapter, we noted that one of the fundamental axes for its development, and probably its main motivation, was in the in-depth study of the periodic (and aperiodic) phenomena described by Fourier analysis. So, it is not surprising that Wiener was very attracted to Fourier analysis. In fact, he had more than close (and very prolonged) contact with it.

Perhaps, as he stated several times, the reason for his interest in Fourier analysis was connected to the need to do valuable work for his colleagues in the electrical engineering department. Moreover, there is also no doubt that Fourier analysis was then (and still is now) one of the central themes of mathematics and one of those with the broadest range of fundamental applications in physics and engineering.

In order to understand the main contributions that Wiener made on this topic, and although we have already provided a brief description of Fourier's analysis, we will now briefly describe the state of the art at the beginning of the last century.

Although Fourier Analysis takes its name from the French mathematician J. B. J. Fourier (1768–1830), we can trace its roots even to the work of Pythagoras. Indeed, the Greeks were aware that the different ways in which a string can vibrate when it is plucked — which are responsible for the sound it emits — follow certain numerical patterns. It is precisely founded on this behavior that the musical scale is defined. Evidently, they did not have the analytical tools essential for an accurate description of an elastic cord's vibrations, but these tools came with time. Indeed, Brook Taylor (1685–1731) proposed in 1715, in his work *Methodus incrementorum direct et inverse*, the vibrating string's problem. It is about determining the movement of an elastic cord as well as its vibration time if it is tensioned by applying an external force and then left free. More precisely, suppose that we have a perfectly taut string of length L, elastic, and attached at the origin of coordinates $(0,0)$ and at the point $(L,0)$, and suppose that we pull it until it reaches the form of the function $y = f(x)$, where of course we have assumed that $f(0) = f(L) = 0$ and $f(x)$ is a continuous function. If we put down the string and let it swing freely, what forms will it take over time? The solution to this fundamental problem obviously passes through two stages. The first one is the derivation of a differential equation whose solutions describe the motion of the string. The second one is the search for these solutions.

Fourier analysis (also called harmonic analysis) deals, in its original form, precisely with the description of periodic phenomena as superpositions of certain simple wave movements. One of the simplest periodic movements is, evidently, one that arises from the study of the oscillations of a spring initially subjected to a certain deformation and then freely abandoned to its whims. This movement, which is called "simple harmonic," is considered simple not only because the differential equation that governs it is easy to obtain (just applying Hook's law) but also because many other periodic movements can be approximated by combining a discrete set of springs. For example, one of the methods initially used to study the vibrating string problem was precisely to idealize the string as the limiting case of a discrete set of beads connected by springs.

Mathematically, the simple harmonic motion is described by the differential equation $x''(t) + \alpha x(t) = 0$, whose solutions are the functions of the form $x(t) = Ae^{i\sqrt{\alpha}t} + Be^{-i\sqrt{\alpha}t}$ (recall that $e^{iwt} = \cos(wt) + i\sin(wt)$). This is why the vibrations described by the functions $\sin(wt), \cos(wt)$ are considered especially important. There is also another fundamental reason that Wiener used to motivate the study of signals as superposition of complex exponentials. It is the following argument: One of the basic principles of deterministic physics (we could also say Newtonian physics) is that nature obeys certain laws and that these laws satisfy two very reasonable hypotheses — they can be expressed through the use of differential (or integral, or integro-differential) equations, and they are also uniform by nature. This means that if a physical law holds at a point in space and a given instant, it should also hold at any other given point and instant: physical laws are therefore necessarily invariant with respect to translations in space and time. Now, it is also true that in the normal practice of physics, laws, if applied under certain simple restrictions, can be approximated by linear operators. Thus, linear and continuous operators that enjoy the additional property of being invariant with respect to translations in time and space are of special interest. These operators are called filters, and it can be proved that complex exponentials are always eigenfunctions of every filter, hence their importance.

Suppose we face a periodic phenomenon and want it to be expressed as a superposition of simple harmonic movements. In that case, it is necessary that the frequencies that appear in the description are all integer multiples of a single fundamental frequency w_0, that (if the period is T) is given by $w_0 = \frac{2\pi}{T}$. This is so because a combination of the form $x(t) = Ae^{iw_1t} + Be^{iw_2t}$ is a periodic function if and only if $\frac{w_1}{w_2} \in \mathbb{Q}$ (so that w_1 and w_2 are both integer multiples of the same frequency w_0). The main consequence of these calculations is that periodic phenomena are described in frequency terms based on a discrete set of numbers: the signal's Fourier coefficients. Specifically, every finite-energy T-periodic signal is uniquely represented as $x(t) = \sum_{k=-\infty}^{+\infty} c_k(x)e^{ikw_0t}$ (where $w_0 = \frac{2\pi}{T}$), simply by taking

$$c_k(x) = \frac{1}{T} \int_0^T x(t)e^{-ikw_0t}dt \text{ for } k \in \mathbb{Z}.$$

These numbers are the so-called Fourier coefficients of the signal, and, obviously, they represent the energy contribution of each of the pure harmonic waves e^{ikw_0t} to the signal. It can be shown that if $x(t)$

is a summable function, then the coefficients $c_k(x)$ tend to zero for $|k| \to \infty$. In particular, the most important energy contribution will always come from the low frequencies. Furthermore, the mapping that sends the signal $x(t)$ to its Fourier coefficients is a Hilbert space isometry. In other words, for the functions $x(t) \in L^2(0, T)$, the following identity is verified:

$$\frac{1}{T} \int_0^T |x(t)|^2 dt = \sum_{k=-\infty}^{+\infty} |c_k(x)|^2.$$

Furthermore, given any sequence of numbers c_k that satisfies $\sum_{k=-\infty}^{+\infty} |c_k|^2 < \infty$, there exists a (unique) signal $x(t) \in L^2(0, T)$ such that $c_k = c_k(x), k = 0, \pm 1, \pm 2, \ldots$.

In a fundamental monograph on Fourier analysis and its applications, which he wrote in 1933, Wiener highlighted in the following words the close link between the Lebesgue integral and Fourier analysis:

> It is known that an adequate theory for the Fourier series can only be established on the basis of Lebesgue integration. All those theorems which proceed from a given function to its Fourier coefficients can indeed be established on the basis of any less inclusive concept, such as that of the Riemann integral, but the fundamental theorem which proceeds from a set of coefficients to the existence of a function having these Fourier coefficients, that of Riesz and Fischer, is simply false for any definition narrower than that of Lebesgue.

To study non-periodic signals $x(t)$, we can perform the following trick: We take a positive value $T > 0$ and consider the signal that results from extending the restriction of $x(t)$ to $[-T, T]$ as a $(2T)$-periodic function, which we denote by $x_T(t)$. Then, we obtain the Fourier series expansion of $x_T(t)$. This series, if it converges, must coincide with $x(t)$ on $[-T, T]$. Finally, we force $T \uparrow \infty$. What happens then? The Fourier coefficients of $x_T(t)$ describe in frequencies the function $x(t)$ for $t \in [-T, T]$ and, furthermore, they are given by the expressions $c_k(x_T) = \frac{1}{2T} \int_{-T}^{T} x(t) e^{\frac{-2\pi k i t}{2T}} dt$. Now, for T big enough and under the hypothesis that $\int_{-\infty}^{\infty} |x(t)| dt < \infty$, we can assume that

$$2T c_k(x_T) \cong \int_{-\infty}^{\infty} x(t) e^{\frac{-\pi k i t}{T}} dt = X\left(\frac{\pi k}{T}\right),$$

where

$$X(\xi) = \int\limits_{-\infty}^{\infty} x(t)e^{-i\xi t}dt$$

is the function that we call the Fourier transform of $x(t)$. Under certain natural conditions, it can be shown that once we know the transform $X(\xi)$, it is possible to recover the initial function $x(t)$.

The aperiodicity of $x(t)$ is then reflected in the fact that, by varying the parameter $T \geq 0$ as well as the integers $k \in \{0, \pm1, \pm2, \ldots\}$, the values $\frac{\pi k}{T}$ run through the entire numerical continuum and, therefore, the description of $x(t)$ in frequency terms requires the use of all the frequencies of the continuum. It was Fourier who, following the calculations we have just outlined, established the following integral formula:

$$x(t) = \frac{1}{2\pi} \int\limits_{-\infty}^{\infty} X(\xi)e^{i\xi t}d\xi.$$

Of course, this formula is valid as long as we impose certain conditions on $x(t)$ and on the way we understand the improper integral. On the other hand, the application that leads $x(t)$ to $X(\xi)$ shares certain formal properties with the Fourier coefficients, which are very useful for applications. Wiener considered especially important the following result, due to Plancherel:

Plancherel's theorem: Assume that $x(t) \in L^2(\mathbb{R})$ is a finite energy signal. Then,

$$X(\xi) = \underset{T\to\infty}{l.i.m} \frac{1}{\sqrt{2\pi}} \int\limits_{-T}^{T} x(t)e^{-i\xi t}dt,$$

where l.i.m denotes the limit in the sense of the norm of $L^2(\mathbb{R})$. Furthermore, $X(\xi) \in L^2(\mathbb{R})$, $x(t) = \underset{T\to\infty}{l.i.m} \frac{1}{\sqrt{2\pi}} \int_{-T}^{T} X(\xi)e^{i\xi t}d\xi$, and

$$\int\limits_{-\infty}^{\infty} |x(t)|^2 dt = \int\limits_{-\infty}^{\infty} |X(\xi)|^2 d\xi.$$

Thus, we can affirm that the fundamental problem of the representation of signals in the frequency domain was solved for periodic signals and for signals of finite energy at the beginning of the 20th century. Wiener would soon increase the class of functions that can be subjected to this type of analysis.

Quantum Mechanics and the Collaboration with Max Born

In 1925, Wiener visited Göttingen, where he had the honor of lecturing at Hilbert's mathematical seminar. His talk was devoted to harmonic analysis and how the uncertainty principle of signal theory results in the physical impossibility of realizing a pure harmonic. Indeed, according to this principle, if the signal $x(t)$ vanishes or is very small (in modulus) outside a certain tiny interval, then its version in the frequency domain, $X(\xi)$, must be quite dispersed (that is, it takes values relatively large or at least not negligible, in the sense that they contribute significantly to the energy of the signal) in a fairly large interval of the real line. Moreover, by changing the roles of $x(t)$ and $X(\xi)$, an analogous statement is satisfied. To measure how scattered a signal may be around the point a, the following magnitude is introduced:

$$\Delta_a x = \frac{\int\limits_{-\infty}^{\infty} |(t-a)x(t)|^2 dt}{\int\limits_{-\infty}^{\infty} |x(t)|^2 dt}.$$

Here, $\Delta_a x$ measures the extent to which the signal $x(t)$ is not concentrated in the neighborhood of the point a.

So, if $x(t)$ has a small modulus outside of a small neighborhood of the point a, then the factor $(t-a)^2$ makes $\int_{-\infty}^{\infty} |(t-a)x(t)|^2 dt$ small compared to $\int_{-\infty}^{\infty} |x(t)|^2 dt$ and therefore $\Delta_a x$ will be very close to zero. On the other hand, if the signal $x(t)$ is scattered (that is, it contains significant information about its energy far from a), then the factor $(t-a)^2$ makes $\int_{-\infty}^{\infty} |(t-a)x(t)|^2 dt$ large compared to $\int_{-\infty}^{\infty} |x(t)|^2 dt$, and therefore $\Delta_a x$ will be a large quantity. The uncertainty principle of signal theory states that it is impossible for signals $x(t)$ and $X(\xi)$ to be simultaneously concentrated around some point on the real line. More precisely, if $a, b \in \mathbb{R}$ and $x(t)$ is a finite energy signal, then

$$(\Delta_a x) \cdot (\Delta_b X) \geq \frac{1}{4}.$$

How can this result be interpreted for acoustic applications? Suppose we want to pluck a guitar string to produce a pure harmonic. Obviously, the signal that we obtain must have a beginning and an end in the time domain because otherwise, it would not be a signal that we can obtain in practice. In particular, in the case of a plucked string, there is some friction with the air, which causes the energy transmitted in the pulse to be dissipated little by little. This means that $x(t)$ will have all the energy concentrated in a finite time interval $[0, T]$. Now, suppose that $x(t)$ behaves roughly like a pure harmonic, so the corresponding signal $X(\xi)$ is concentrated around a certain point $w_0 \in \mathbb{R}$ and therefore $\Delta_{w_0} X$ must be a quantity very close to zero. Well, in this case, the uncertainty principle guarantees that $\Delta_a x$ must be large for every $a \in \mathbb{R}$ (since $(\Delta_a x) \cdot (\Delta_{w_0} X) \geq \frac{1}{4}$ and $\Delta_{w_0} X$ is very small). This contradicts that the signal $x(t)$ vanishes outside of the interval $[0, T]$. In conclusion, it is impossible to approximate the behavior of a pure harmonic in practice. The reason we can perceive music is because our ears are unable to perceive the disturbances in the frequency domain produced by the fact that sounds begin and end. It seems that the ear only perceives the "stationary" part of the signal, which can behave as a pure harmonic. Another consequence that can be deduced from the uncertainty principle, and that Wiener liked very much, is that for a piece of music to sound good, bass sounds (with small frequencies) must be prolonged in time longer than treble ones. Concretely, the product of the time duration times the frequency of the sound should necessarily be greater than a certain positive constant. If we do not follow this rule, then we will hear little else but noise.

Of course, even though we have used the word "principle" to refer to the uncertainty result, the truth is that we are dealing with an authentic mathematical theorem. It is not a physics hypothesis or an axiom from which we wish to obtain new results, but rather a valid result for all finite energy signals. Two fundamental details that we should highlight at this point are the following: First, the uncertainty principle can be established — making the appropriate conventions — for functions of several variables. Second, finite energy signals (in one and several variables) are indeed an excellent model for treating different physical problems. Thus, depending on the situation we are modeling, the signals $x(t)$ and $X(\xi)$ have specific physical meanings, and the uncertainty principle is applied with a special meaning to each model. Wiener was fully aware of this. In his 1925 lecture, he presented several examples of physical interest where a small variation in the values of a certain physical magnitude necessarily produces a disturbance in another complementary magnitude's values. His examples were related to optics and Maxwell's theory of heat.

When Wiener delivered his lecture, Heisenberg and Born were engaged in a discussion on the foundations of quantum mechanics (of which Heisenberg, with the introduction of his matrix mechanics, would give the first precise mathematical formulation in 1925). When they tried to measure the position and momentum on an atomic scale, it seemed that classical physics laws did not apply because these magnitudes seemed to follow a strange complementary pattern. Understanding why this happened was a fundamental problem they wanted to solve. Obviously, the topic chosen by Wiener for his lecture aroused great interest and, in fact, this sudden interest motivated Wiener to get down to work, study some quantum mechanics, and — during the academic year of 1925 — take advantage of Born's visit to MIT to investigate with him the mathematical foundations of that discipline.

Indeed, after the introduction by Heisenberg of a formalism suitable for the treatment of quantum mechanics, but which was somewhat obscure in its physical motivations and in its mathematics — which involved a non-commutative product, among other things — Born, whose training in mathematics was much more robust than Heisenberg's, had found that the rules he was using had a relatively simple expression if they were written in terms of matrices. This pleased Heisenberg but not many other physicists, who saw the new quantum theory as something artificial and difficult to handle. Born was not satisfied with the precise formulation that he had achieved and, as the good mathematician that he was, he tried to generalize Heisenberg's "discrete" theory (which only took into account the possibility that the involved particle system acquired a discrete set of energy levels) to a "continuous" theory in which the spectrum could take arbitrary values. It was precisely in this attempt that Born and Wiener would collaborate, publishing an important article in 1926. In their paper, they substituted Heisenberg's matrices for certain operators defined on a space of functions. Thus, for the first time, the idea was raised that a physical quantity can be represented as an operator defined on a certain space of functions.

Norbert Wiener and Max Born (1882–1970).
Courtesy: MIT Museum.

Unfortunately, the underlying function space (which, they suspected, should at least contain the class of Besicovitch quasi-periodic functions that we will introduce later) was not defined with enough precision in the work of Born and Wiener. This was undoubtedly considered a weak point in their research and was probably why his article was not received in its true value either then or later.

In his description of his contact with Born, Wiener commented that

> When Professor Born came to the United States he was enormously excited about the new basis Heisenberg had just given for the quantum theory of the atom. This theory was an essentially discrete one, and the tools for its study consisted of certain square arrays of numbers known as matrices. The separateness of the lines and columns of these matrices was associated with the separateness of the radiation lines in the spectrum of an atom. However, not all parts of the spectrum of an atom are made of discrete lines, and Born wanted a theory which would generalize these matrices or grids of numbers into something with a continuity comparable to that of the continuous part of the spectrum. The job was a highly technical one, and he counted on me for aid.
>
> (...) I had the generalization of matrices already at hand in the form of what is known as operators. Born had a good many qualms about the soundness of my method and kept wondering if Hilbert would approve of my mathematics. Hilbert did, in fact, approve of it, and operators have since remained

an essential part of quantum theory. They were introduced about the same time by the independent work of Paul Dirac in England. Moreover, they turned out to be useful in tying up another form of quantum mechanics just being invented in Vienna by Erwin Schrödinger with the Heisenberg form of the theory.

Thus, we can affirm that Wiener felt in a certain way that he was the "father" of one of the fundamental ideas of new physics. However, it seems that the pressure under which they worked on this topic — where, in Wiener's words, "some young people like Heisenberg, Dirac, Wolfgang Pauli and John von Neumann were making discoveries almost daily" — led Wiener to abandon this research topic as soon as Born finished his stay at MIT, as he felt that "he was not able to contribute fresh ideas." Still, Wiener liked working with Born, whom he described as

A calm, gentle, musical soul, whose chief enthusiasm in life has been to play two-piano music with his wife.

That same year, physicists were surprised by the creation of a new mathematical formalism for quantum mechanics, this time at the hands of E. Schrödinger (1887–1961), who gave a description of the discipline founded on an idea proposed in 1924 by Louis de Broglie (1892–1987): matter waves. The surprise was that the two theories, Heisenberg's matrix mechanics and Schrödinger's wave mechanics, had a very different mathematical (and physical) appearance. Yet, all the examples studied produced exactly the same results and predictions. Although the fundamental problem of the matrix mechanics was the diagonalization of an (infinite) matrix, in wave mechanics, the basic problem consisted of solving a certain partial differential equation. How was it possible that both theories explained quantum phenomena equally well? The only option, on which Schrödinger himself immediately set to work, was for the two to be mathematically equivalent. Schrödinger believed he could demonstrate the equivalence, but he only succeeded in proving that wave mechanics is part of matrix mechanics (something that was evidently to the advantage of his greatest opponent, Heisenberg). Dirac and Jordan took the next step, but the final touch, the final completion of the puzzle, came from the hands of J. von Neumann, who published in 1932 a monograph entitled *Mathematical Foundations of Quantum*

Mechanics. In his book, Hilbert space formalism was used to give an axiomatization of quantum mechanics in which both theories, the matrix and the wave, fit. Von Neumann's formalization used the ideas that Wiener and Born had anticipated a few years earlier, but this time the space of functions on which the operators that define the states of a quantum system were defined was clearly fixed: It was just a separable Hilbert space. The unification of the theories of Heisenberg and Schrödinger was based on an important theorem of functional analysis due to Riesz and Fischer, and which implies, among other things, that all separable Hilbert spaces are isometric. In particular, the space $\ell^2(\mathbb{Z})$ of matrix mechanics is indistinguishable from the space $L^2(\mathbb{R})$ of wave mechanics.

For his part, Heisenberg would demonstrate in 1927 that from the postulates of matrix mechanics an uncertainty principle could be deduced, with a formulation analogous to the principle discussed by Wiener in his 1925 lecture but with a profound physical meaning.

Concretely, he proved that $(\Delta q) \cdot (\Delta p) \geq \frac{h}{4\pi}$, where $h = 6.6 \times 10^{-34}$ J is Planck's constant, q denotes the observed position, and p is the measured moment of a particle. This inequality means that it is impossible to locate any particle fully. In particular, it forces us to think about any particle's position and momentum in probabilistic terms since it is impossible to know both variables simultaneously with absolute precision.

Interestingly, in his recollections of the period when he worked with Born on quantum mechanics, Wiener included some philosophical comments on twentieth-century physics's primary goals, highlighting the absence of a unifying theory that consistently included quantum mechanics, electromagnetism, and relativistic gravitational theory. Indeed, this is a subject on which many leading physicists still work today — without great progress. On this topic, Wiener believed that

> The modern physicist is a quantum theorist on Monday, Wednesday, and Friday and a student of gravitational relativity theory on Tuesday, Thursday, and Saturday. On Sunday the physicist is neither but is praying to his God that someone, preferably himself, will find the reconciliation between these two views.

Heaviside's Operational Calculus, Causal Operators, and Distributions

Wiener was assigned a very complicated task from the electrical engineering department: It was necessary to mathematically justify some of the ideas and methodologies that were common among engineers but were not fully understood. At the forefront of these problems was explaining Heaviside's operational calculus.

Oliver Heaviside (1850–1925) was an important English scientist with a particular interest in electricity and the applications of mathematics in engineering. He is often considered self-taught as he only received formal education until he was sixteen, at which point he dropped out of school and went to work in a telegraph company. At the age of 24, Heaviside left this job and, for the next 35 years, devoted himself exclusively to research in electrical engineering and mathematics. During that time, he was financially dependent on his family and friends, which did not prevent him from saying what he thought, getting into complex controversies, and even insulting those who paid his bills. Of course, he remained single. He never had social skills and was even annoyed at others' mere presence. If he met a neighbor on one of his walks in the countryside, it was common for him to write a note in his diary referring to this event as something particularly unpleasant. In short, he had a devilish character, a lot of pride, and little sense of humor. Moreover, his scientific writings lacked the usual formalism and contained acid criticisms of other scientists' work and opinions, all of them better positioned than he. With all these ingredients, it is comprehensible that he was involved in several tough controversies throughout his life. Thus, for example, he discussed with the highest authority in England on telegraph matters, Sir G. Preece, how things were being done for the proper functioning of the great submarine cable that stretched between the United Kingdom and the United States. Furthermore, after introducing his operational calculus, he also had important differences with the Cambridge mathematicians, who viewed Heaviside's strange manipulations with a critical eye. With geologists, he discussed the problem of accurately determining the age of the earth. With Bell Laboratories (that is, the research department of the telephone and telegraph company AT&T), he had problems with several patents that they created using the ideas of Heaviside, thanks to which the company was able to establish long-distance telephone connections that soon competed with the telegraph.

In his book, *Invention. The Care and Feeding of Ideas*, Wiener devoted an entire chapter to describe Heaviside's life and the conflict he had with Bell Labs. There, he introduced him as follows:

Oliver Heaviside
(1850–1925).
Courtesy:
Public domain,
Wikimedia Commons.

> Heaviside was born poor, lived poor, and died poor; he was sincere, courageous, and incorruptible. In addition, he could use the very limited mathematics available to the electrical engineers of his day with an unorthodox skill that was to disconcert a whole generation of mathematicians.

Wiener admired Heaviside, so the proposal he received from MIT's electrical engineering department had to be very challenging to him.

Let us see, by showing an example, what is operational calculus. Suppose we want to solve the differential equation associated with an RC circuit, which is the equation given by $RCy'(t) + y(t) = x(t)$, where R, C are constants, $x(t) = i(t)u(t)$ is the potential difference applied to the circuit, and is $y(t)$ the output of the system.

We have set $x(t) = i(t)u(t)$, where $u(t) = \begin{cases} 1 \ (t \geq 0) \\ 0 \ (t < 0) \end{cases}$ denotes Heaviside's unit step function, because the circuit is assumed to be inactive before and active from $t = 0$ on. To keep things simple, let us assume the input current is constant, $i(t) = E$, and see how the circuit acts.

If we denote by p the operator $p = \frac{d}{dt}$ and by p^{-1} the operator $p^{-1}f = \int_0^t f$, then the above equation can be formally written as $(RCp + 1)y(t) = x(t)$, so we can rewrite it as $y(t) = \frac{1}{RCp+1}x(t)$. Now,

$$\frac{1}{RCp + 1} = \frac{1}{RC}\frac{1}{p}\frac{1}{1+\frac{1}{RCp}} = \frac{1}{RC}\frac{1}{p}\sum_{k=0}^{\infty}(-1)^k\frac{1}{(RC)^k}\frac{1}{p^k}$$

$$= \sum_{k=0}^{\infty}(-1)^k\frac{1}{(RC)^{k+1}}\frac{1}{p^{k+1}}.$$

On the other hand, it is clear that $\frac{1}{p}u(t) = \int_0^t ds = tu(t)$, $\frac{1}{p^2}u(t) = \int_0^t s\,ds = \frac{t^2}{2}u(t)$, and

$$\frac{1}{p^k}u(t) = \frac{t^k}{k!}u(t), \quad k = 1, 2, \dots .$$

Hence,

$$y(t) = E\sum_{k=0}^{\infty}(-1)^k \frac{1}{(RC)^{k+1}} \frac{1}{p^{k+1}} u(t)$$

$$= (-1)E\sum_{k=0}^{\infty}(-1)^{k+1}\frac{1}{(RC)^{k+1}}\frac{t^{k+1}u(t)}{(k+1)!} = E(1 - e^{-t/(RC)})u(t),$$

which is the correct solution to the differential equation. To check that indeed $y(t) = E(1 - e^{-t/(RC)})u(t)$ is a solution to the equation, it is enough to distinguish the cases $t \le 0$ (where everything is trivial) and $t > 0$ (where $y(t) = E(1 - e^{-t/(RC)})$)). Then, we assume that $t > 0$, take the derivative $y'(t) = E\frac{1}{RC}e^{-t/(RC)}$, and substitute the result in the original equation to obtain that $RCy'(t) + y(t) = Ee^{-t/(RC)} + E(1 - e^{-t/(RC)}) = E$, which is what we wanted to prove.

Obviously, these computations are not justified as we have exposed them here. For example, it is not clear what the meaning of the expression $\frac{1}{p}$ may be, nor do we know what meaning could be given to the convergence of the involved series. It is really surprising that the above computations produce the correct solution to the problem. Yet they do. So, there is an interesting question: Why? How is it that we can operate algebraically with the operator p and its "inverse" and, finally, use a power series expansion of $\frac{1}{p}$ to obtain a series of operators that, once applied to the input $x(t)$, will give us the solution of the problem? Is it possible that such a large accumulation of inaccuracies and abuses of language can nevertheless lead us to a victorious exit?

It is clear that such a strange treatment of the differential and integral operators could not be admitted in the mathematical circles of the time and all the more so when these calculations were presented in a period in which mathematicians were particularly concerned with rigor. Following the invention, a few centuries ago, by Newton and Leibniz, of the infinitesimal calculus, there had been an almost unlimited (and, of course, unprecedented) expansion of mathematical analysis. In a mad rush to describe the physical phenomena that govern the world, mathematicians and physicists were carried away by the enthusiasm. They were not very careful in some of their deductions, especially in the topic of differential equations (both ordinary and partial). This had the effect that, precisely at the end of the 19th century, there was an intense crisis in the foundations of mathematical analysis, a crisis that was related to the fact that there are functionals that do not reach their minimum value on any specific function — something that had been taken for granted in some applications of variational calculus. From there, the attempt to solve the difficulties led to transferring the difficulty to increasingly elementary subjects, even reaching the very concept of a number. It must be said that, although at first glance this would seem scandalous and difficult to fix, it is nothing more than the consequence of the fact that in mathematics we move from the simplest and most intuitive concepts (as one might think that the concept of real number is) towards the more difficult and abrupt ones. Therefore, a failure above can only come from the very foundation of mathematics, which is in many ways a "slippery floor" on which one has to tread carefully. Thus, when mathematicians of the stature of Weierstrass had just launched themselves into a detailed study of the foundations of analysis, it could not be considered the ideal moment for Heaviside's "intuitions."

In addition to the power series of $\frac{1}{p}$ just introduced, Heaviside encountered examples of physical problems for which the application of his technique required the use of rational powers of p. But what can we intuit to be $p^{\frac{1}{2}}$? Heaviside himself would recognize in his writings that such an operator was

> ... unintelligible from the point of view of ordinary differentiation ...

Of course, this did not prevent him, to his colleagues' surprise and chagrin, from working with such objects. Furthermore, Heaviside dared to use divergent series in his computations, and this was the step that overwhelmed the patience of Cambridge's mathematicians. Heaviside was dismissive of the numerous criticisms he had received from them — who, among other feats, managed to interrupt a series of articles he was publishing in the *Proceedings of the Royal Society.*

After all, and despite the fact that he had recently been appointed as Fellow of the Royal Society, he was an electro technician. He had the soul of an engineer. What really worried him was knowing that his methods produced correct solutions. Explaining in depth why they worked was, of course, the task of mathematicians. He proposed the problem, and the problem was important.

We cannot say that Heaviside introduced operators in mathematics. On the contrary, they had been used systematically by mathematicians of the stature of Lagrange, Laplace, Fourier, Cauchy, and Boole, among others. What Heaviside did was instead an "illicit use" of operators. Despite how well assimilated they already were in mathematics, he introduced a strong controversy around operators, even if the "operational" methods had been used long before.

The idea of treating differential operators as algebraic entities seems to have originated from the fact that the notation introduced by Leibniz for the derivative of a function, $f' = \frac{df}{dx}$, gave rise to treating "differentials" df, dx as "separate" entities that could be algebraically manipulated. Even today, it is common in the first-year calculus courses to use these "tricks" to make changes of variables in the computation of integrals, among other things. As Wiener comments as soon as he begins his 1926 article,

> The operational calculus owes its inception to Leibniz, who was struck by the resemblance between the formula for the n-fold differentiation of a product and the formula for the nth power of a sum. Lagrange carried this analogy further, and set up a regular algorithm in which the differential operator plays the part of a sort of *hypercomplex* unit. It is to him, for example, that we owe the symbolic formula
>
> $$f(x + h) = \left(e^{h \frac{d}{dx}} \right) (f(x)).$$
>
> This operational calculus was further developed by many mathematicians, including Laplace and, more especially, Boole, who employed it effectively in the well-known theory of the linear differential equation with constant coefficients.

Thus, these "operational" methods had been used before Heaviside by some of the most important mathematicians of the nineteenth century. Among them, it is worth highlighting P. S. Laplace (1749–1827) and A. L. Cauchy (1789–1857), but we can also include J. L. Lagrange (1736–1813) (who tried to algebrize the entire differential calculus) and L. Arbogast (1759–1803). However, from the first third of the 19th century onwards, operational methods fell into disrepair, and Heaviside was precisely responsible for relaunching them, focusing especially on differential equations that appear in electrical circuits. Obviously, as his methods produced the correct solution to numerous engineering problems, they became popular with engineers. Little by little, the need to provide a rigorous explanation of them became apparent.

At MIT, the person who prompted Wiener to study operational methods was D. C. Jackson (1865–1951), who in 1920 was the head of the electrical engineering department. Wiener and Jackson knew each other from childhood. According to Wiener, Jackson had searched for some time — without success — for an engineer with a special sensitivity for mathematics or a mathematician with engineering skills. This person should help him solve some of the difficulties that existed to give a solid foundation to the new electronic engineering. Of course, Jackson found the right person in Wiener and would soon ask him the questions that had haunted him for many years.

Electrical and Electronic Engineering

In the early 20th century, electrical engineers typically distinguished between two distinct trends: "strong currents" engineering, or power engineering, and "weak currents" engineering. Power engineering was particularly concerned with electric generators, motors, transformers, etc., as well as the problem of large-scale power distribution. This subject was quite well established in 1920 and currently resides in what we call electrical engineering. On the other hand, weak current engineering, that is, electronic engineering or communications engineering, was still taking its first steps. Obviously, the telephone's appearance had been a fundamental impulse for the study of this new branch of engineering. After all, the telephone was known to work by transmitting small, highly fluctuating electrical impulses, and therefore it was essential to understand how such signals behave. Fourier analysis would have much to contribute in this direction, and Norbert Wiener would soon become one of the greatest exponents on this topic.

Of course, before Wiener, there were already several mathematicians, physicists, and engineers who had tried to give a solid foundation to operational calculus. Laplace had used his one-sided transform,

$$\mathcal{L}[f](s) = F(s) = \int_0^\infty f(t)e^{-st}dt,$$

coupled with the fact that

$$\mathcal{L}(f^{(n)})(s) = s^n F(s) - (s^{n-1}f(0) + s^{n-2}f'(0) + \cdots$$
$$+ sf^{(n-2)}(0) + f^{(n-1)}(0))$$

to solve a given differential equation of the type

$$a_0 y(t) + a_1 y'(t) + \cdots + a_n y^{(n)}(t) = x(t).$$

Apply the transform to both sides of the equality to obtain that

$$(a_0 + a_1 s + \cdots + a_n s^n)Y(s) + (b_0 + b_1 s + \cdots + b_{n-1}s^{n-1}) = X(s)$$

for a certain polynomial $b_0 + b_1 s + \cdots + b_{n-1}s^{n-1}$. In this way, we transform the problem of searching for a solution of the differential equation in calculating the inverse one-sided Laplace transform of the signal

$$Y(s) = \frac{X(s) - (b_0 + b_1 s + \cdots + b_{n-1}s^{n-1})}{a_0 + a_1 s + \cdots + a_n s^n}.$$

For his part, Cauchy had used similar techniques but based on the Fourier transform. The problem with Laplace and Cauchy's justifications was that they imposed by force that both the inputs and the outputs of the physical systems involved had a particular behavior at infinity. In both cases, the functions involved had to converge to zero at infinity with a certain speed. However, Heaviside used his methods in more general cases than these. Could there be a better method to justify his calculations?

In Wiener's words,

> Although Heaviside's developments have not been justified by
> the present state of the purely mathematical theory of oper-
> ators, there is a great deal of what we may call experimental
> evidence for their validity, and they are very valuable to the
> electrical engineer. There are cases, however, where they lead to
> ambiguous or [even] contradictory results. It has hence become
> important to put them on a sound mathematical basis, or, fail-
> ing that, to establish heuristically criteria for the avoidance of
> contradiction. V. Bush has given a set of formal criteria of this
> type ...

Wiener would use Fourier analysis methods, together with an
adequate interpretation of the asymptotic expressions involved, to
develop his approximation to operational calculus. The idea he had,
and which was also central to a good part of his future contributions, was to broaden the class of signals to which Fourier methods can be applied (something that, as we will see in the next
section of this chapter, was not easy to foresee and also required a
great imagination and a good intuition about the associated physical
processes).

In the operational calculus, functions $f(t)$ are taken for which a
power series expansion $f(t) = \sum_{k=0}^{\infty} a_k t^k$ is known and the associated operators $f(\frac{d}{dt}) = \sum_{k=0}^{\infty} a_k \frac{d^k}{dt^k}$ are defined. One of the keys to
Heaviside's theory is that if we take $g(t) = e^{iwt}$ and $f(t) = e^{i\alpha t}$, then
the operator defined by

$$
L = g\left(\frac{d}{dt}\right) = \sum_{k=0}^{\infty} \frac{(iw)^k}{k!} \frac{d^k}{dt^k}
$$

verifies that

$$
L(f)(t) = f(iw)f(t).
$$

To check it, just keep in mind that for this specific function the relation $\frac{d^k}{dt^k}(f)(t) = f^{(k)}(t) = (i\alpha)^k f(t)$ holds, and make the corresponding substitution in the definition of the operator. Wiener thought
that this property is fundamental to the justification of the operational calculus and, referring to it, commented:

When applied to the function $f(t) = e^{int}$, the operator $f\left(\frac{d}{dt}\right)$ is equivalent to the multiplier $f(ni)$. The result of applying a given operator to a given Fourier integral may naturally be conceived, then, as the multiplication of each e^{int} term in the integral by a multiplier depending only on n. That is, the operator $\frac{d}{dt}$ has no particular location in the complex plane but ranges up and down the entire axis of imaginaries. It is thus, too much to expect in general that any particular series expansion or other analytic representation of f should make $f\left(\frac{d}{dt}\right)$ converge when applied to a purely arbitrary function. In the present paper, the device is adopted of dissecting a function into a finite or infinite number of ranges of frequency and of applying to each range the particular expansion of an operator yielding results convergent over that range. It is by this method of dissection that Heaviside's asymptotic series are justified.

With this approach, it is evident that Wiener would try to find a frequency representation (that is, one dependent on pure harmonics) for functions that are as arbitrary as possible. Classical Fourier analysis had already solved the cases of periodic functions and finite energy functions. However, Wiener wanted to apply the method to aperiodic functions that did not have to decay to zero at infinity. It was precisely this concern that led him to establish, little by little, a new "generalized" Fourier analysis, which he would later develop in great detail. Specifically, in the article we are commenting on, he performs his calculations for functions satisfying that they and their squares are locally integrable and whose growth at infinity is of polynomial order.

However, Wiener's article on Heaviside's operational calculus, released in 1926, was important for other reasons as well. In it, Wiener introduces the concept of causal system (although he uses the terminology "retrospective system") and, in addition, provides a first approximation — 20 years ahead of time — to the current theory of distributions. Finally, he solves in the distributional sense one of the most important equations of the moment, the telegraph equation.

Causal systems are those whose output at the instant of time t depends exclusively on the past values (that is, those before t) of the system's input. These systems are very important in physics and telecommunications. Keep in mind that these systems are precisely those that can be performed in real time. Wiener would not limit himself to introducing this concept but, as we will see later, he would establish the fundamental principles by which they are governed, and, in addition, he would deal with the problem of their physical realization "in the metal." Interestingly, the mathematics involved in this theory necessarily went through the use of functions of a complex variable and gave rise to some exciting "pure mathematics" problems.

Distributions are a fairly radical generalization of ordinary functions. The first to use distributions were the physicists, who saw in them an intuitive way of dealing with phenomena involving certain types of sudden changes, so much so that their time duration is very small compared to the measurement times allowed by the physical system where they occur. For example, this phenomenon occurs in an electrical circuit subjected to certain "specific" electrical impulses. They also take place in quantum mechanics. Indeed, it was P. A. M. Dirac (1902–1984) who, in 1926, introduced a suitable mathematical notation for the treatment of these impulses, although without being able to provide a deep justification for it. Specifically, Dirac assumed the existence of a certain function $\delta(t)$ with the property that $\int_{-\infty}^{+\infty} \delta(t)dt = 1$, despite the fact that $\delta(t) = 0$ as soon as $t \neq 0$. The idea is that the function $\delta(t)$ concentrates all its energy at the moment $t = 0$ (for this, $\delta(0) = +\infty$ is required). Thus, $\delta(t)$ could represent an isolated electrical impulse that introduces a unit of energy into the system.

By 1930, physicists had adopted Dirac's intuitive formalism to such an extent that it was clear that it would not be easy to eliminate it from their calculations, so its mathematical formalization was obviously a pending and urgent task.

Obviously, from the point of view of classical mathematics, $\delta(t)$ cannot be a function in the usual sense of the word. So, what is it? Physicists and mathematicians alike soon agreed that the new objects, which they would call generalized functions or distributions, were nothing but "functionals."

The underlying idea is actually very simple. Let's imagine that we want to characterize the number 3. Wouldn't it be possible to do this (in a probably far-fetched but correct way) by giving a precise description of how the "multiply by three" operator works? In other words, it is clear that the application $L : \mathbb{R} \to \mathbb{R}$ given by $L(x) = 3x$ has all the information necessary to characterize the number 3. Then, when we want to identify a function $x(t)$ instead of a number, we could propose that its characterization can be achieved based on the operator $L_x(\varphi) = \int_{-\infty}^{\infty} x(s)\varphi(s)ds$. Obviously, to proceed that way it is necessary to set the class of functions with which we want to work and, in terms of this, define a class of functions $\varphi(t)$ that has the property of being wide enough to allow the operators so defined to distinguish any pair of different functions. This means that the set of functions $\varphi(t)$ must be such that if $L_x(\varphi) = L_y(\varphi)$ for every function $\varphi(t)$, then we will have that $x(t)$ and $y(t)$ must coincide (functions $\varphi(t)$ are called test functions and they are always required to vanish at infinity).

This is precisely the way in which the theory of distributions works. The interesting thing is that once the class of functions $x(t)$ to be identified and the class of test functions are established, if we look at the linear and continuous functionals defined on the space of test functions, we immediately detect that some of these functionals do not come from any ordinary function $x(t)$. Therefore, if we interpret the functionals we are considering as a reasonable generalization of ordinary functions, we have constructed a certain new class of "functions," the so-called distributions.

A fundamental aspect of the theory of distributions is that it allows taking as test functions functions $\varphi(t)$ that are differentiable as many times as one wishes (this is an approximation property) and, therefore, if we take into account that in the case that the signal $x(t)$ has a derivative in the usual sense of the term, a simple process of integration by parts (that is, integrating taking into account the rule for the derivative of the product of two functions and the fundamental theorem of calculus) leads to the relation

$$L_{x'}(\varphi) = \int_{-\infty}^{\infty} x'(s)\varphi(s)ds = (-1) \int_{-\infty}^{\infty} x(s)\varphi'(s)ds.$$

It follows that, once we have identified the signals x with the functionals L, we can naturally define the derivative L' of the functional L as the functional

$$L'(\varphi) = (-1)L(\varphi').$$

Thanks to this idea, we can differentiate every distribution as many times as we want.

Wiener's introduction of the concept of distribution (although he did not use this terminology) gives a clear sample of his genius. Wiener was interested in providing some reasonable method for the solution of certain partial differential equations that appeared in electrical engineering and that had the unpleasant characteristic that, viewed as linear systems, they typically allowed discontinuous inputs. An important example is the behavior of telegraph signals.

The telegraph is based precisely on the transmission of certain electrical impulses of different durations, which, moreover, are discontinuous by definition. Thus, if we want to know what type of signal is transformed into a succession of impulses (long or short) that we introduce at one end of the cable when they have traveled hundreds or thousands of kilometers, we will have to resort to the differential equation of the telegraph, which is (in the case that the line does not have losses because it is perfectly insulated) the equation given by $v_{xx} = RCv_t + LCv_{tt}$, where $v(x,t)$ denotes the voltage of the cable at the instant t and at the position x, and R, C, L are, respectively, the resistance, capacity, and inductance per unit of cable length, which are assumed to be constant.

To be well posed, the problem requires forcing certain initial and boundary conditions. We assume that these are given by $v(x, 0) = 0$ and $v(0, t) = f(t)$ for a certain function $f(t)$ that we interpret as the input of the system (the output will be the function $g(t) = v(l, t)$, if the cable has length l). Heaviside introduced this differential equation in contrast to Thomson's model (Lord Kelvin) for his studies of the transoceanic submarine cable. The equation studied by Thomson did not take inductance into consideration, so that it is obtained from the previous one by taking $L = 0$. Interestingly, this gives us the same equation that appeared in Fourier's work on the diffusion of heat in a metal rod. Heaviside derived the telegraph equation from Maxwell's equations.

Telegraph versus Telephone

In his studies on the telegraph, Heaviside found a way to solve one of the fundamental problems associated with transmission lines (and, in particular, telephone transmission): the problem of distortion of transmitted waves. Let us see how J. A. Fleming (1849–1945) describes this problem in his monograph on the history of electricity:

> Unfortunately, it happens that electric waves of different wavelengths do not travel at the same speed through the cable and they are not equally attenuated. Shorter waves travel faster and attenuate faster. Therefore, the waveform is distorted by the line or cable and, beyond a certain limit distance, this distortion deprives the articulated sound of any quality or character that gives it its meaning. The problem presented to early electricians was: how to prevent or remedy this wave distortion?

The ways used to solve this critical difficulty before the arrival of Heaviside were two: reduce the cable's resistance or reduce the electrical capacitance of the line. Neither of these things was reasonable from a practical point of view. For example, to reduce resistance, it is necessary to use copper or bronze cables, but these are very expensive. On the other hand, gutta-percha, which was the only insulator available for the construction of submarine cables, had poor properties if what we were looking for was to reduce the line's capacitance. After solving his telegraph equation using operational techniques, Heaviside warned — although he was not heard — that the solution was to increase the line's inductance, something that is feasible if we wrap the conductive wire with an iron wire that runs through it following a spiral curve along its entire length (uniform load) or inserting into the circuit every certain distance a coil of copper wire insulated and wound to a bundle of iron wire (load with coils). Heaviside's research was later used by AT&T in the United States to patent its new long-distance transmission lines. It turns out that the telephone, in its early formats, had not been conceived for the use of the general public and, much less, for long-distance use. Instead, it was thought of as a valuable device for commercial communication within the same city. In no case was it intended to replace the telegraph. However, if we take into account the calculations developed by Heaviside and the fact that it is relatively easy to increase the inductance in a transmission line, the bet on a telephone that would compete with the telegraph was served. According to Wiener's version, the problem that AT&T executives had was that Heaviside's work was too old to be able to draw a commercial use patent from it. And without a patent, there were no guarantees that they could monopolize the market and, therefore, the investment required was "high risk." As a solution, they looked for an independent engineer (a certain Pupin). Working together with another engineer from the company (Campbell), they could locate a weak point in Heaviside's calculations. Thus, just by improving the fringes, a patent could be created. The aspect that Campbell and Pupin improved precisely explained the proper distribution of the loading coils. With this, they created the patent, and the company paid Pupin almost half a million dollars.

AT&T engineers were angry with Pupin and with company executives for two main reasons. First, it seems that the added scientific work had been done mainly by Campbell, who would not charge anything for the patent since he was an employee of AT&T, and, second, in the patents, they had ignored Heaviside. Most of the engineers there felt they had to do something for Heaviside (who was not financially well off). There was so much pressure that the company tried to pay Heaviside, but Heaviside — who was stubborn as a mule — flatly refused to receive any compensation unless the patent was awarded to him. For his part, Pupin tried to clear his conscience by writing a book that he entitled *From immigrant to inventor*. The book became a bestseller in the United States, as it was considered the model story of personal triumph, the realization of the American dream of equal rights and opportunities. Pupin would receive the Pulitzer Prize for this work in 1924.

However, in his account of the events, Pupin clearly underestimated Heaviside's contributions and, although it turned out to be very stimulating for the country's schoolchildren (who came to have it as required reading), AT&T engineers who knew the situation "up close" were not at all satisfied with the result.

Wiener recounts the Heaviside conflict in his book *Invention* (published posthumously):

> Corporations have no feelings to be hurt, nor does a man of easily moved feelings rise to high office in one, but there was a man who was hurt and hurt greatly. While Heaviside had the sardonic satisfaction of incorruptibility, Pupin was in a really difficult situation. He had accepted half a million dollars which was legally his for an invention which he had only made in a very Pickwickian sense.
>
> (...) The fact is that circumstances had joined the story of Prometheus with the story of Marlowe's Dr. Faustus. Heaviside may have been a very snuffy lower-middle-class Prometheus, but at least he had snatched a piece of fire for mankind. If the vultures of poverty and the sense of persecution were gnawing at his liver, he shared with Prometheus the sense of having performed a godlike feat. Pupin, on the other hand, had wrapped his soul as part and parcel of a commercial bargain. When a soul is bought by anyone, the devil is the ultimate consumer. Even a public penance was denied Pupin.
>
> Although he was unable to contain himself in silence, the lies and bluster to which he was forced to resort must have echoed hollowly in the empty space where his soul had been.

It must be said, however, that there is another (more plausible) version of this story, according to which there was no conflict between Heaviside and AT&T. Instead, it was between Heaviside and the Post Office, which was responsible for the telegraph service in England, as well as between Heaviside and Pupin. Campbell and Pupin worked independently, but Pupin, who was not under the orders of any company, was able to make his patent freely, which forced AT&T to pay for it (see [Kr]).

Wiener became obsessed with Heaviside's life, probably because he too had his difficulties with Bell Labs, so much so that, in August 1957, he abandoned the writing of his essay on the invention and began to write a novel, which he titled The Tempter, in which he tells of the life of two scientists trapped by industry and the temptations which they have to face, forced to choose between maintaining their dignity or achieving personal success.

If one wants to solve it in the ordinary sense, the telegraph equation requires solutions that are — at least — twice differentiable. How are we going to achieve such a solution if the input to the system is, in fact, a function that clearly has jump discontinuities? A similar situation had been occurring in other physical contexts, such as the vibrating string problem. It is straightforward to press an elastic cord so that its initial shape is a function with spikes, but that does not prevent the desired movement from taking place when it is released, nor does the motion fail to be governed by the same differential equation.

Wiener, in response to this situation, in his 1926 article, stated that:

> It is a matter of some interest, therefore, to render precise the manner in which a non-differentiable function may satisfy in a generalized sense a differential equation.

This would be nothing more than the expression of desire or frustration of many mathematicians and physicists who constantly faced unpleasant situations such as the presence of spikes in a vibrating string, if it were not for the explanation of a method that would later be the basis for the creation of the theory of distributions. Evidently, the vibrating string problem had given rise to an intense discussion, part of which was precisely to explain in what sense a function with peaks could produce solutions to a differential equation. Thus, Wiener continued his reflection as follows:

> Let $G(x,y)$ be function positive and infinitely differentiable within a certain bounded polygonal region R of the XY plane,

vanishing with its derivatives of all orders at the periphery of R and zero outside R. Then there exists a function $G_1(x,y)$ such that

$$\iint\limits_{R} (Au_{xx} + Bu_{xy} + Cu_{yy} + Du_x + Eu_y + Fu)G(x,y)dxdy$$

$$= \iint\limits_{R} u(x,y)G_1(x,y)dxdy$$

for all u with bounded summable derivatives of the first two orders, as we may show by integration by parts. Thus a necessary and sufficient condition for u to satisfy our differential equation almost everywhere is that

$$\iint\limits_{R} u(x,y)G_1(x,y)dxdy = 0$$

for every possible G (as the G's form a complete set over any region), and that u has the required derivatives. We can thus regard a function orthogonal to every G_1 as satisfying our differential equation in a generalized sense.

And then, he used the operational techniques that he had just introduced to solve, in this generalized sense, the telegraph's equation. It is unnecessary to say that Wiener's generalized solutions are nothing else but the current weak solutions or solutions in the distributional sense of the equation under consideration.

Mikusinski's Operational Calculus

Jan Mikusinski (1913–1987). Courtesy: Public domain, Wikimedia Commons.

In 1943, the Polish mathematician Jan Mikusinski (1913–1987) introduced an algebraic formulation of Heaviside's operational calculus. The story around this contribution is interesting. The first publication of Mikusinski's original article only had seven copies that he photocopied himself (using an X-ray camera) and that he distributed among those attending one of the mathematical seminars organized in secret by Professor Tadeusz Wazewsky (1896–1972) during the Nazi occupation of Poland.

The article would see the light of day when it was published in 1944 by the well-known journal *Studia Mathematika*. However, the work was not signed with the full name of the author, but he just put the initials J. G. M. (of which it seems that the "G" stood for "Genius," and was an acronym that Mikusinski would use for a certain time). It was the beginning of an important research work carried out by Mikusinski and many of his students. Mikusinski is also recognized for his introduction of distribution theory through a technique based on the convergence of sequences, which is much easier to approach than the topological method (based on functional analysis) proposed by Schwartz.

As is clear from the computations explained in this section, the idea behind Heaviside's operational calculus is that perhaps there is an algebraic field K that contains the operator $p = \frac{d}{dt}$, in the sense that if $\cdot$ denotes the product of K, then $p \cdot x = \frac{dx}{dt}$. If this is so, as $p \neq 0$, we can speak of its algebraic inverse and, as $1 = p \cdot p^{-1}$, we have $x(t) = \frac{d(p^{-1} \cdot x)}{dt}$. Therefore (except for the intervention of a constant, which we will assume by agreement is equal to zero), the relation $(p^{-1} \cdot x)(t) = \int_0^t x(s)ds$ holds.

Indeed, Mikusinski proved that such a field can be built. To formalize this construction in a consistent mathematical theory, it was, however, necessary to demonstrate some complicated technical results. In particular, it was required to prove that if the convolution of two signals $x(t), y(t)$ defined for $t \geq 0$, which is given by the expression

$$(x * y)(t) = \int_0^t x(s)y(t - s)ds,$$

results in the identically null signal, then one of the signals must also be identically equal to zero.

This result had been proved by E. C. Titchmarsch (1899–1963) in 1926, and it is certainly not easy to deduce. (In 1979, the Japanese mathematician S. Okamoto would find a simple way to deduce Mikusinski's operational calculus avoiding this technical lemma.) Mikusinski defines the product of two functions as their convolution. With the objective that, when the new concept is applied to any pair of complex numbers, the result should coincide with the usual product, we are going to introduce the product in a slightly different way. Specifically, we define, for functions defined in $[0, \infty)$, which are differentiable and whose derivative is locally integrable, the product

$$(x \cdot y)(t) = \frac{d}{dt} \int_0^t x(s)y(t-s)ds.$$

It is easy to show that the set M of the functions thus defined is closed for the product that we have just introduced and, in fact, with this product and with the ordinary sum of functions, forms a ring. On the other hand, if

$$(x \cdot y) = 0,$$

then $\int_0^t x(s)y(t-s)ds$ is a constant c. Now, the substitution $t = 0$ shows that this constant is necessarily $c = 0$ and therefore we can apply the Titchmarsch theorem to conclude that $x = 0$ or $y = 0$.

It follows that M is an integral domain and, therefore, we can construct its field of fractions K.

If we use the definition of derivative as a limit and the fact that for every continuous function $x(t)$ we necessarily have that $\int_a^b x(s)ds = (b-a)x(\xi)$ for a certain value $\xi \in (a, b)$, it is very easy to check that

$$(x \cdot y)(t) = \frac{d}{dt} \int_0^t x(s)y(t-s)ds = y(0)x(t) + \int_0^t x(s)y'(t-s)ds.$$

Let us now consider the function $I(t) = t$. Applying the above formula with $y(t) = I(t)$, we get that $(x \cdot I)(t) = \int_0^t x(s)ds$. Consequently, if we use the notation $p = \frac{1}{I}$, then $(p \cdot x)(t) = x(0)p + x'(t)$. Thus, if we identify p with the operator "multiply by p" (that is, the functional that sends x to $(p \cdot x)(t) = x(0)p + x'(t)$), and we apply it exclusively to functions that vanish at $t = 0$, then p is nothing else but the differential operator $p = \frac{d}{dt}$, which is what we were looking for.

It is imperative to note that the initial set M, formed by differentiable functions with locally integrable derivatives, is exempt from restrictions on the decay to zero at infinity. Thus, although the justification of operational methods based on the use of Laplace's transform eventually prevailed, only with the advent of Mikusinski's operational calculus was a definitive answer to the problems posed by Heaviside achieved. In particular, it was an answer in which it was unnecessary to assume a priori any particular behavior of the involved functions near to infinity.

Vannevar Bush (1890–1974).
Courtesy: MIT Museum.

In addition to his article for *Math. Annalen*, in 1926, Wiener collaborated with Vannevar Bush, one of the promising youngsters in MIT's electrical engineering department, contributing with an appendix to a book on linear systems that he was writing. Bush would fondly recall his work with Wiener, going so far as to say that never before had he imagined that a collaboration with a mathematician could be not only fruitful and intense but also enjoyable and fun. Like Wiener, Bush had studied at Tufts College but had graduated a couple of years later. Starting in 1927, and for at least a decade, Wiener would maintain ongoing contact with Bush, whom he would periodically advise on the design of several analog machines for the calculus of integrals and the solution of differential equations. Although Wiener was very clumsy from a practical point of view (he would never manage to build the simplest artifact by himself), his understanding of engineering problems was exceptionally profound, and, of course, his advice was profitable. On his part, Bush was a genuine "handyman." In modern engineering history, he represents one of the last (and most skilled) members of the old school. He ended up getting involved in politics, and, as we shall see, his participation in the United States defense program during the Second World War was crucial. Subsequently, he received the Medal of Merit from Truman and the National Medal of Science from Johnson. He also received (in 1943) the Edison Medal for his contributions to engineering and, in particular, to the applications of mathematics in electrical engineering. The year 1926 would be a decisive year for the advancement of science. While Wiener was spinning the idea of a generalized solution for partial differential equations, several physicists founded quantum mechanics. Nothing would ever be the same in mathematics or, of course, in physics.

Chapter 4

Climbing Mathematical Olympus

Generalized Harmonic Analysis

Wiener decided to extend the basic results of Fourier analysis to the study of aperiodic signals that do not decay to zero at infinity. These signals appeared in his justification of the operational calculus, but they represented some well-known physical phenomena, such as white light. Indeed, as on so many other occasions, Wiener found the motivation to study the new harmonic analysis in physics.

In a seminal article, entitled "Generalized Harmonic Analysis," published in *Acta Mathematica* in 1930, Wiener commented that the tools of classical Fourier analysis "are not suitable for the study of a ray of white light that is supposed to remain for a (undefined) long time" and added:

> Nevertheless, the physicists who first were faced with the problem of analyzing white light into its components had to employ one or the other of these tools [(i.e.: Fourier series and Fourier transforms)]. Gouy accordingly represented white light by a Fourier series, the period of which he allowed to grow without limit, and by focusing his attention on the average values of the energies concerned, he was able to arrive at results in agreement with the experiments. Lord Rayleigh on the other hand, achieved much the same purpose by using the Fourier integral, and what we now should call Plancherel's theorem. In both cases one is astonished by the skill with which the authors use clumsy and unsuitable tools to obtain the right results, and one is led to admire the unfailing heuristic insight of the true physicist.

These comments summarized the fundamental motivation with which Wiener dared to propose that the old Fourier Analysis needed a significant extension if it wanted to be useful in the study of physical phenomena that implied the use of signals that, without being periodic, were not losing energy over time. Furthermore, these phenomena were more frequent than might appear at first glance. For example, the signals transmitted by a telephone line or the graphs drawn by an encephalogram clearly fell within this concept. Wiener was deeply convinced that these functions should be susceptible to harmonic analysis (and therefore describable in spectral terms). These signals' spectra had to reflect an intermediate state between the discrete spectrum of the Fourier series and the continuous spectrum of the Fourier transform. However, how to delineate the functions that should be treatable with this new analysis? It would be best to see what others had done, for which Wiener studied the works of the British physicists Lord Kelvin (1824–1907), Lord Rayleigh (1842–1919), Sir Arthur Schuster (1851–1934), and Sir Geoffrey Taylor (1886–1975).

The class considered by Wiener is the set S of the functions $f\colon \mathbb{R} \to \mathbb{C}$ that are measurable in the Lebesgue sense and have the property that their associated covariance function, which is the function φ defined by

$$\varphi(t) = \lim_{T\to\infty} \frac{1}{2T} \int_{-T}^{T} f(s+t)\overline{f(s)}\,ds,$$

exists for every $t\in\mathbb{R}$ and is continuous everywhere. This set of functions is pervasive. The signals observed in all physical processes that we have mentioned previously, precisely those that Wiener had in mind, belong to this set. Not only that, S contains a class of functions whose study was booming at the beginning of the last century, namely, the quasi-periodic functions of H. Bohr (1887–1951) and A. S. Besicovitch (1891–1970). Furthermore, the sample functions of numerous random processes (including the Brownian motion) also belong to S. This made the study of this set extremely interesting to Wiener's eyes. If he managed to transfer the basic results of Fourier analysis to this "space" of functions, he would obtain a resounding success, and his international recognition would be fixed in the history of mathematics.

The Quasi-Periodic Functions of Bohr and Besicovitch

One way to include Fourier analysis as a particular case of a broader theory is to propose a concept that generalizes the idea of periodicity and try to express the new functions as a superposition of complex exponentials. This is precisely what H. Bohr (1887–1951), the younger brother of the well-known physicist N. Bohr (1885–1962), did by introducing the definition of quasi-periodic function.

Suppose we have a function $x(t)$ defined for all real values of t and let $\varepsilon > 0$. We say that the real number L is a translation number associated with ε if $\sup_{t \in \mathbb{R}} |x(t) - x(t+L)| \leq \varepsilon$. Now, we will say that $x(t)$ is quasi-periodic if for each $\varepsilon > 0$ there exists a number $L(\varepsilon)$ such that every interval of the form $(a, a + L(\varepsilon))$ contains a translation number associated with ε.

Bohr showed two exciting results on quasi-periodic functions. In the first one, he characterizes them as those functions that can be uniformly approximated on the real line with linear combinations of complex exponentials that are not harmonically related (that is, whose angular frequencies are not all integer multiples of a given frequency). The second result is an adapted version of Parseval's Theorem valid for quasi-periodic functions. Specifically, it states the following:

Theorem. If $x(t)$ is a quasi-periodic function, then the limit

$$\lim_{T \to \infty} \frac{1}{2T} \int_{-T}^{T} x(t) e^{-i\lambda t} dt$$

exists for every $\lambda \in \mathbb{R}$, being non-zero for at most a countable set of parameter values λ. Furthermore, if these values are those described by the sequence $\{\lambda_k\}_{k=1}^{\infty}$, then, taking

$$A_k = \lim_{T \to \infty} \frac{1}{2T} \int_{-T}^{T} x(t) e^{-i\lambda_k t} dt,$$

we have that

$$\lim_{n \to \infty} \lim_{T \to \infty} \frac{1}{2T} \int_{-T}^{T} \left| x(t) - \sum_{k=1}^{n} A_k e^{i\lambda_k t} \right|^2 dt = 0.$$

Although Bohr proved these results before Wiener (who was the second to give a proof) and there were also other proofs by H. Weyl (1885–1955) and De la Vallée Poussin (1866–1962), Wiener was delighted with his own proof because, of all of them, it was the only one compatible with a unified treatment of quasi-periodic functions and functions with a continuous spectrum.

From a mathematical point of view, the best way to motivate the study of the covariance function $\varphi(t)$ is to start with a general trigonometric polynomial $f(t) = \sum_{k=0}^{n} a_k e^{i\lambda_k t}$, and to obtain that its covariance function $\varphi(t) = \lim_{T \to \infty} \frac{1}{2T} \int_{-T}^{T} f(s+t)\overline{f(s)}ds$ is given by $\varphi(t) = \sum_{k=0}^{n} |a_k|^2 e^{i\lambda_k t}$. This means that the covariance function $\varphi(t)$ preserves all the information about the amplitude of the spectrum of $f(t)$ and therefore it preserves all the information about the energy contributed by each pure harmonic to the spectrum of $f(t)$, while eliminating the information regarding the phases of that spectrum. It is not difficult to show that $\varphi(t)$ is continuous everywhere if and only if it is continuous at the origin of coordinates $t = 0$ since it verifies the following inequality:

$$|\varphi(t) - \varphi(t+\varepsilon)| \leq \sqrt{[2\varphi(0) - 2Re(\varphi(\varepsilon))]\,\varphi(0)}.$$

Obviously, $\varphi(0) = \lim_{T \to \infty} \frac{1}{2T} \int_{-T}^{T} |f(s)|^2 ds$ can be interpreted as the power of the signal $f(t)$. Now, if we try to identify what part of the power lies in the frequency interval $[-A, A]$ and we make A tend to infinity, we will obtain that the limit to consider is precisely $\lim_{\varepsilon \to 0} \varphi(\varepsilon)$. Furthermore, this limit is necessarily less than or equal to $\varphi(0)$. Thus, if the covariance function $\varphi(t)$ is not continuous at the origin, this can only be interpreted as the existence of hidden frequencies (infinite, said Wiener) that contribute energy to the signal $f(t)$, which is an undesirable situation. This is why Wiener demands that

$\varphi(t)$ be continuous at zero (and, henceforth, throughout all the real line) to cope with a generalized harmonic analysis of $f(t)$.

In order to properly speak of a new Fourier analysis, Wiener needed to prove a version of Plancherel's Theorem for the class of functions S. This means that a valid spectrum concept had to be found for such functions. Furthermore, it was necessary to construct a transform that would relate each function $f(t)$ of S with its spectrum, and this transform should be invertible and preserve the signal's power.

The first step for such a construction was to prove that if $f(t)$ belongs to S, then $\int_{-\infty}^{\infty} \frac{|f(t)|^2}{1+t^2} dt < \infty$, and therefore the existence (in the sense of the energy norm) of the limit

$$W(f)(\xi) = l.i.m_{A \to \infty} \left[\frac{1}{\sqrt{2\pi}} \left[\int_{-A}^{-1} + \int_{1}^{A} \right] \frac{f(t)e^{-i\xi t}}{-it} dt \right.$$
$$\left. + \frac{1}{\sqrt{2\pi}} \int_{-1}^{1} \frac{f(t)(1 - e^{-i\xi t})}{it} dt \right]$$

can be guaranteed. This function $W(f)(\xi)$ is very important because, in the case that the function had finite energy, and therefore admits a Fourier transform $F(\xi)$, it is not difficult to verify that $W(f)(\xi)$ would be a primitive of $F(\xi)$ (that is, $\frac{d}{d\xi}(W(f)(\xi)) = W(f)'(\xi) = F(\xi)$). Wiener showed that there is a close link between the covariance function and the new function $W(f)(\xi)$.

Specifically, he proved that

$$\varphi(t) = l.i.m_{\varepsilon \to 0} \frac{1}{4\pi\varepsilon} \int_{-\infty}^{\infty} |W(f)(w + \varepsilon) - W(f)(w - \varepsilon)|^2 e^{iwt} dw$$

and as a consequence, there exists a non-negative, monotonic, non-decreasing, and bounded-variation function $\Lambda(\xi)$ such that $\varphi(t) = \int_{-\infty}^{\infty} e^{it\xi} d\Lambda(\xi)$. Furthermore, the function $\Lambda(\xi)$ can be retrieved from the covariance function thanks to the formula

$$\lim_{A\to\infty} \frac{1}{\sqrt{2\pi}} \left[\int_{-A}^{-1} + \int_{1}^{A} \right] \frac{\varphi(t)e^{-i\xi t}}{-it} dt + \frac{1}{\sqrt{2\pi}} \int_{-1}^{1} \frac{\varphi(t)(1 - e^{-i\xi t})}{it} dt$$

$$= \Lambda(\xi) + cte.$$

Finally, if we choose $\Lambda(\xi)$ having the property that $\lim_{\xi\to-\infty} \Lambda(\xi) = 0$, then $\Lambda(\xi)$ represents the total power in the spectrum of $f(t)$ for the frequencies between $-\infty$ and ξ. The function $\Lambda(\xi)$ was called — following the terminology used by Sir A. Schuster (1851–1934) — the integrated spectrum (or periodogram) of $f(t)$ by Wiener. There are very diverse forms that $\Lambda(\xi)$ can take, and Wiener devoted considerable effort to studying them, giving examples of discrete, continuous, and mixed spectra.

J. Benedetto (born in 1939) reformulated these results in the 1970s in terms of distributions. Specifically, it can be stated that if $f(t)$ satisfies the inequality $\int_{-\infty}^{\infty} \frac{|f(t)|^2}{1+t^2} dt < \infty$, then its Wiener transform $W(f)(\xi)$ exists and it satisfies, in the distributional sense, the identity $W(f)' = F$, where F denotes the Fourier transform of f in the distributional sense.

Once the transform $W(f)(\xi)$ was introduced, Wiener proved for it a result analogous to Plancherel's Theorem. Namely, he proved that

$$\lim_{T\to\infty} \frac{1}{2T} \int_{-T}^{T} |f(t)|^2 \, dt = \lim_{h\to 0} \frac{1}{4\pi h} \int_{-\infty}^{\infty} |W(f)(\xi + h) - W(f)(\xi - h)|^2 \, d\xi$$

for the functions $f(t)$ of $\boldsymbol{S}$. Thus, his aim to find a tool analogous to classical harmonic analysis that would be valid for a wide class of non-periodic and infinite energy signals was fully satisfied.

This result, which in the specialized literature is called the Wiener–Plancherel theorem, was by no means easy to achieve. Indeed, as early as 1926, Wiener had a statement of it, but his proof depended on the hypothesis that every positive function g verifies the identity

$$\lim_{T\to\infty} \frac{1}{T} \int_{-T}^{T} g(t)dt = \lim_{h\to 0} \frac{2}{\pi h} \int_{0}^{\infty} g(t)\frac{\sin^2(ht)}{t^2} dt,$$

something that cannot easily be shown. Precisely the search for a demonstration that the previous identity is indeed true led him to a field of study that, at that time, was new to him but in which he managed to leave an indelible mark: the Tauberian theorems. The fact is that Wiener finally managed to prove this identity and, with it, lay the foundations for a new harmonic analysis applicable to phenomena as diverse as Brownian motion, fluctuating signals from telephone lines, and light.

Statistical Aspects of Generalized Harmonic Analysis: Ergodic Theory

Wiener's new harmonic analysis became an instrumental theory in physics and engineering, different from everything previously known, because it allowed working with random phenomena, especially with certain stochastic processes that naturally appear in communications theory. Let us see how this link is established.

In all the calculations that we explained in the previous section, the signals involved were functions in the ordinary sense of the term. That is, they were functions that depended in an exact way on the only variable under consideration, time. This type of dependency is deterministic. Given an instant of time t_0, the value $f(t_0)$ that the signal takes at that instant is perfectly known. On the other hand, we know that there are many physical processes for which we cannot guarantee a concrete deterministic dependence between the variables involved. If we were to carry out the same experiment several times under the same physical conditions, we could not guarantee the same result. When this happens, we say that we are in the presence of a random phenomenon. If we consider the result of physical experiments of this type for each instant of time, what we obtain is a stochastic process $\{X_t\}_{t \in \mathbb{R}}$ (see Chapter 2 to recall the corresponding definitions). For convenience, we are now going to denote these processes as if they were ordinary functions. Thus, in this section, the expression $f(t_0)$ represents the value of the random variable X_t for $t = t_0$ and the expression $f(t)$ represents a sample function of the process $\{X_t\}_{t \in \mathbb{R}}$. The important thing about the stochastic process under consideration here is that the different random variables X_t share certain probabilistic properties. For example, suppose the process represents the noise term produced in a transmission line. In that case, we know that all the random variables X_t have the same probability distribution function. The characteristics of this distribution depend on the type of noise considered and the line's physical characteristics. Then, for each sample function of the process, we can consider its covariance function

$$\varphi(t) = \lim_{T \to \infty} \frac{1}{2T} \int\limits_{-T}^{T} f(s+t)\overline{f(s)}ds.$$

Since the covariance function of a signal $f(t)$ removes all the information related to the signal's phase spectrum, we can take advantage of this situation to guarantee a vast collection of signals that share the same covariance function.

Wiener's idea was to assume that we are dealing with a stochastic process with the important property that all its sample functions have the same covariance function. This is something similar to assuming, for example, that we study a random variable with a given type of distribution (e.g., normally distributed) or to fix some value related to the random variable such as, for example, its expectation, its standard deviation, or some other moments. We can accept Wiener's hypothesis as good if we assume that our stochastic process's sample functions are produced in the same way by the same source and under identical physical conditions. As the covariance function determines the spectrum or integrated periodogram of any of the sample functions under consideration, what we are doing translates into the hypothesis that this spectrum is a distinctive feature of the system.

However, it must be taken into account that his approach entails some difficulties of a practical nature. One of the most important is that the covariance function's analytical determination implies knowing a priori the signal in its entirety (past and future) since it is defined as a temporal average that covers the entire existence of the signal. Another difficulty, typical of the mathematical models used in communication theory and in physics, is that an infinite duration of the signal is assumed, which is unrealistic. In practical cases, we only know the past of the signal $f(t)$. Furthermore, there is no possibility of reliably predicting its future since it is a stochastic process's sample function.

Wiener then raised the fundamental question of whether there are reasonable hypotheses to guarantee equality between the covariance function $\varphi(t)$ and the temporal average that we can compute, given by $\lim_{T \to +\infty} \frac{1}{T} \int_{-T}^{0} \overline{f(s)} f(t+s) ds$ (that is, we know the values of $f(t)$ previous to the origin, and we have imposed $f(t) = 0$ for $t > 0$).

To address this problem, he took the Gibbs statistical mechanics model and the related ergodic theory. The underlying idea is that, in special circumstances, it is possible to exchange a time average (such as the average velocity or the average position of a particle) for certain "set" averages that take into account the behavior of an entire family of particles at a given moment (or signals, if we are in a communication theory context). The set averages can be compared to the average behavior of each of these particles (signals). We will formulate these conditions with some precision both for the case of statistical mechanics — which is where they arose — and for the theory of communication.

In statistical mechanics, a system of particles is considered (for example, an ideal gas) that consists of a finite number (let's say N) of moving particles and which we consider as an isolated system. In other words, we assume that there are no more interactions than those that these N particles may have with each other. A complete description of the system at a given instant will therefore require knowing a total of $3N$ spatial coordinates — with every three numbers, we would be indicating the position of one of its component particles — and another $3N$ coordinates that would indicate the moments corresponding to each of the particles. That is, a point in a $6N$-dimensional space would determine the complete description of the system at a given instant, and the description of the behavior of the system over time would be represented by a curve of points in that space, which physicists call the space of phases or phase space of the system.

According to Newtonian mechanics, each phase space's point would uniquely determine a single curve in it. However, we know that it is impossible to simultaneously measure the position and momentum of a particle with infinite precision. Furthermore, when dealing with macroscopic situations, the number of particles involved is obviously too large for calculations. In fact, the problem of determining the behavior of a group of only three particles — not to say N — from their positions and moments at a given instant (knowing the

mass of each particle) is still one of the most important open problems in mathematical physics. Thus, it would be reasonable for our
interest to shift from the study of the trajectories described by each
of the particles in the system to the study of the system's mean
behavior.

This means that we are not interested in the specific trajectory
carried out by each particle but instead in some macroscopic properties such as the particles' average speed. The idea that Gibbs had
to address these questions is very simple: Instead of assuming that
we know the exact distribution of the particles of the system in its
initial position, we assume that it follows a certain probability distribution and from there we allow the use of Newton's deterministic
laws. Under these assumptions, Gibbs needed, for the development
of his theory, to assume that if the energy of the system is constant
(which is a consequence of assuming that this is an isolated system),
then certain temporal averages taken on the trajectory described
by each of the particles of the system can be expressed as an average obtained by integrating over the submanifold of the phase space
that is obtained by considering precisely those phases that are compatible with the energy of the system. This sub-variety is commonly
known as the ergodic surface. In an attempt to justify Gibbs' results,
Maxwell introduced the ergodic hypothesis, which consists of assuming that if we let the system evolve freely over time, sooner or later,
it will pass through each and every one of the points of the ergodic
surface. However, although the ergodic hypothesis is very natural
from a physics point of view, from a mathematical point of view, it
turns out to be always false because it is impossible to find a differentiable curve that fills a plane, a space, or any other variety of
higher dimension.

The consequence of this negative result was the replacement of
the ergodic hypothesis by a weaker version, called the quasi-ergodic
hypothesis, which consists of assuming that there are isolated systems with the property that if we let them move freely and choose an
arbitrary point of the ergodic surface, then, for whatever precision
level we set in advance, there will come a time when the point in the
phase space that describes the state of the system is as close to the
given point as we want. This hypothesis was admissible from a mathematical perspective since it is relatively easy to find smooth curves

whose trajectory is dense in a variety of dimensions greater than or equal to two. Still, it took some work to show that, indeed, there are systems verifying the quasi-ergodic hypothesis. This work was successfully completed by G. D. Birkhoff and von Neumann (among others) thirty years after the statement of this hypothesis. Wiener also made an important contribution to this matter by publishing in 1939 an article entitled "The Ergodic Theorem" in the *Duke Math. Journal* in which he presented a unified view of the results that were known so far, with new demonstrations of them, as well as some interesting generalizations. In another article of great importance, which he would publish in the *American Journal of Mathematics* in 1938 under the title "Homogeneous Chaos," he would deal with the old problem of turbulence, which had resisted him years ago but with which he was now able to deal, thanks to the ergodic techniques he was developing.

In terms of communication theory, the quasi-ergodic hypothesis translates into assuming that we are working with a stationary family of random functions (which we interpret as messages, as noise, etc.) produced by sources of the same nature and with a continuous range of possible values, with the property that the amplitudes reached over time by any of the messages of the family describe a dense subset of the range of possible values. Therefore, fixing any value within that range and a number ε arbitrarily small, there are instants of time for which our signal's value is at a distance smaller than ε from the given value.

Wiener was able to demonstrate, based on the results of ergodic theory, that if the stochastic process $\{X_t\}_{t\in\mathbb{R}}$, whose sample functions are precisely the family of messages selected by us, is ergodic and satisfies a certain symmetry property, then the chain of equalities,

$$\varphi(\tau) = \lim_{T\to+\infty} \frac{1}{T} \int\limits_{-T}^{0} f(t)f(t+\tau)dt = \int\limits_{-\infty}^{\infty} \int\limits_{-\infty}^{\infty} xy P_{X_t X_{t+\tau}}(x,y,\tau)dxdy,$$

is satisfied, where $P_{X_t X_{t+\tau}}(x,y,\tau)$ denotes the second probability density function of the stochastic process and $f(t)$ is any of its sample functions. Under these conditions, it is then possible to calculate the spectrum associated with the stochastic process thanks to its own probabilistic description and the exclusive knowledge of the past of any of its sample functions. It is in this way that, again, Wiener solved an essentially practical problem — the determination of the covariance function $\varphi(t)$ from the past of $f(t)$, with important applications in engineering — by using a complicated mathematical theory, the ergodic theory, with profound roots in measure theory, statistical mechanics, and Fourier analysis.

One Year in Cambridge

A new edition of the International Congress of Mathematicians would take place in Zurich in 1932. Wiener was eager to participate, so he planned to spend the 1931–1932 academic year at Cambridge and attend the congress as a representative of MIT. Sure enough, he got the necessary permits and funding. He traveled to Cambridge with the whole family, where he would be welcomed as a visiting professor. Just before the boat trip to England, both Wiener and his youngest daughter Peggy, who was then only one and a half years old, fell ill with flu. Norbert quickly recovered, but his daughter got worse over the days, and as soon as they arrived in London, she had to be hospitalized with bronchopneumonia. Fortunately, the girl recovered in few days, and the entire family was able to move to Cambridge, where they settled in a "typical middle-class English house" near the countryside.

At the end of fall, Hardy, then a professor at Cambridge, proposed that Wiener teach a course on the Fourier transform and write a monograph based on that course that Cambridge University Press would publish. Wiener was delighted to accept. Moreover, knowing that his classes were taught on behalf of Hardy, following a tradition typical of that university, he felt very flattered by the offer, which made him see himself during that year as an "unofficial professor" at Cambridge. Indeed, Wiener taught his course and wrote an important monograph entitled *The Fourier Integral and Some of Its Applications* in which, among other things, he summarized his contributions to general harmonic analysis. Furthermore, the book would include a beautiful proof of Plancherel's theorem, based on the Fourier transform's description as an operator of the space $L^2(\mathbb{R})$ into itself, whose eigenfunctions are the Hermite functions, which form an orthogonal basis of $L^2(\mathbb{R})$, and whose associated eigenvalues are $1, -1, i, -i$.

During his year at Cambridge, Wiener had numerous scientific contacts in Cambridge, Oxford, London, and continental Europe. For example, he received invitations from W. Blaschke (1885–1962) in Hamburg, K. Menger (1902–1985) in Vienna, and P. Frank (1884–1966) in Prague. Later, he would remember some of these names with some sadness:

> Merely to think of these names brings to my mind the vicissitudes which these men have suffered since. Blaschke became, during the Second World War, if not an ardent Nazi, at least an ardent supporter of the Nazis, and he wrote articles in ridicule of American mathematics. In particular, he showed his contempt for the mathematical school of Princeton, which he called "a little Negro village." Menger later on came to the United States as a refugee. I helped him find a position at Notre Dame. I believe he is now in the Illinois Institute of Technology. Frank also came over as a refugee from Hitler and has recently retired from Harvard.

In addition to his classes and writing his monograph, Wiener also took advantage of his stay from a personal point of view by establishing a friendship with the well-known English biologist and philosopher J. B. S. Haldane (1892–1964), one of the first scientific popularizers of that country.

Haldane frequently visited the Cambridge Philosophical Society library. It was there that Wiener first read his science fiction story *The Gold Makers* and, realizing that the photo on the back cover of the book was precisely that of his library mate, he steeled himself and presented the writer with an observation about the name of one of the novel's characters. This is how a friendship that would last a lifetime was born between the two. In his memoirs, Wiener would comment:

> I have never met a man with better conversation or more varied knowledge than J. B. S. Haldane.

J. B. S. Haldane
(1892–1964)
with Marcello
Siniscalco (1924–2013)
(standing) in
Andhra Pradesh,
India, 1964.
Courtesy: Dsiniscalco,
Public domain,
Wikimedia Commons.

Haldane would also influence Wiener's philosophical thought and, quite possibly, his political ideas (Haldane was a left-wing man). Nevertheless, the most important thing is that Wiener found in him a friend, with whom he would enjoy not only conversing but also taking long walks or even bathing in the river:

> We continued to see a lot the Haldanes, and I used to go swimming with him in a stretch of the River Cam, which passed by his lawn. Haldane used to take his pipe in swimming. Following his example, I smoked a cigar and, as has always been my habit, wore my glasses. We must have appeared to boaters on the river like a couple of great water animals, a long and a short walrus, let us say, bobbing up and down in the stream.

First Results on Wave Filters: Paley and Lee; The Lee–Wiener Machine

Wiener was very interested in wave filters. One of his greatest successes, which he would remember fondly for the rest of his life, was precisely the characterization he arrived at in 1933, with the help of the English mathematician R. E. A. C. Paley (1907–1933), of the physically realizable filters. These filters are straightforward to characterize in the time domain because they are precisely those that use only a limited part of the input signal's future (so that it is enough to wait a finite time to provide the filter response in any instant). Nevertheless, the characterization in terms of the filter transfer function was indeed a complicated problem.

Wiener, who had met Paley in London during his year in Cambridge, met him again at the Zurich congress and proposed that he visit MIT in the following academic year. This was done (Paley got a grant from the Common Wealth). There, they worked together on issues related to the use of the Fourier transform in the complex plane. They spent long hours arguing in front of a huge blackboard and, each time it filled with formulas, one of the two sat down and "copied the most relevant" before erasing and continuing with the calculations. Wiener had a beautiful memory of those days. He admired Paley's ability to solve challenging questions.

Most of the problems they dealt with were proposed by Wiener and generally motivated by physics and engineering, but Paley did not seem concerned with applications. Still, Wiener remembered with admiration how Paley tackled problems with an arsenal of surprising tricks, and they were "falling" one after another. In his autobiography, he recalls with emotion how they came to the characterization of physically realizable filters:

> One interesting problem which we attacked together was that of the conditions restricting the Fourier transform of a function vanishing on the half line. This is a sound mathematical problem on its own merits, and Paley attacked it with vigor, but what helped me and did not help Paley was that it is essentially a problem in electrical engineering. It had been known for many years that there is a certain limitation on the harpness with which an electric wave filter cuts a frequency band off, but the physicists and engineers had been quite unaware of the deep mathematical grounds for these limitations. In solving what was for Paley a beautiful and difficult chess problem, completely contained within itself, I showed at the same time that the limitations under which the electrical engineers were working were precisely those which prevent the future from influencing the past.

However, those days of hard work ended in tragedy. Paley, who not only was an excellent mathematician but an impetuous fan of climbing and skiing (hobbies he shared with Littlewood, Hardy's inseparable collaborator, whose skill in mathematics he also admired), decided to take a few days off and go skiing with some friends from Boston in the Canadian Rockies.

The group's guide had warned that it was forbidden to ski in certain areas because there was a serious danger of avalanches at that time (it was April). Now, as Wiener would later assert,

> To forbid a thing to Paley was to make quite sure that he would do it.

So, Paley disobeyed the order and got caught in an avalanche of snow. From there, they sent a telegram to Wiener informing him of the death. Wiener had to report what happened to the deceased's mother and his English colleagues. All of this plunged him into a depressed state from which it took him some time to recover.

R. E. A. C. Paley
(1907–1933).
Courtesy:
Public domain,
Wikimedia Commons.

Despite the brevity of the contact between Paley and Wiener, their names remained almost inseparable for posterity because some of the results that they jointly demonstrated have become part of the foundations of harmonic analysis in the complex domain. This effect was reinforced because in 1933, after receiving the Bôcher Prize for his research in Tauberian Theorems, which we will talk about later, Wiener was invited by the American Mathematical Society to write a monograph for the collection of the Society's "Colloquia" and he decided to write for that volume his results with Paley (jointly with his own results in the subject) on the Fourier transform in the complex domain. Wiener had the deference to put Paley as co-author of the text, even though Paley had already passed away when Wiener was invited to write the monograph. The volume, titled *Fourier Transforms in the Complex Domain*, appeared in 1934 and was an immediate success. It was the best tribute Wiener could pay to his young friend.

The Basic Theory of Wave Filters

Wave filters were one of the fundamental ingredients in Wiener's work on communications engineering (and later on cybernetics). Thus, it is perhaps worth briefly explaining the foundations of that theory.

A filter is just an operator defined between two signal spaces (functions, distributions, etc.) with the properties of being linear, continuous, and invariant by translations. This last property means that the effect produced on the system's output by a delay in the input signal is simply the same delay for the output signal. If the complex exponential signal $e_w(t) = e^{iwt}$ is taken as input of the filter L, then it is easy to prove that $L(e_w(t)) = \lambda(w)e_w(t)$ for a certain complex number $\lambda(w)$.

This means that the operator L has a straightforward behavior on the signals that we can represent as a superposition of complex exponentials. Therefore, it is one of the most natural motivations that we can give to study Fourier analysis in its diverse variants.

Depending on the nature of the signal spaces involved, the filters will have a particular analytical expression. In the simplest case, when the input signals form a Hilbert space, and the output signals are bounded signals, the filters are always convolution operators. That is, they are operators of the form $L(x)(t) = \int_{-\infty}^{\infty} x(s)h(t-s)ds$, in the analog case, or of the form $L(x)[n] = \sum_{k=-\infty}^{\infty} x[k]h[n-k]$, in the discrete-time case. This representation makes it very easy to characterize the causality or physical realizability of wave filters in terms of the unit impulse response h, in the time domain. To do this, it is enough to bear in mind that if we want the filter to be causal, $L(x)$ can only depend on the past values of the input x, and therefore h must vanish for the negative values of its variable. Physical realizability translates into using at most a limited portion of the future of the input, which is guaranteed by the fact that h vanishes for all values less than a given one.

If we take the Fourier transform on both sides of the equality $L(x)(t) = \int_{-\infty}^{\infty} x(s)h(t-s)ds$, the filter L is characterized in the frequency domain as an operator of the form $L(X)(w) = X(w)H(w)$, where $X(w), H(w)$ represent the Fourier transforms of $x(t), h(t)$, respectively (an analogous property is given for the discrete-time case). The "transfer function" $H(w)$, therefore, determines the filter.

What's more, this function gives a very intuitive description of the filter's action on the different frequency components of the system's input signal. Thus, the electrical engineer's most natural action is to design his filters in the frequency domain precisely by giving an adequate description of the associated transfer functions. However, this raises a question that is not easy to answer: Given a function $H(w)$, how do we know if it corresponds to the transfer function of a causal filter or, at least, of a physically realizable filter?

It is precisely this question that Paley and Wiener answered in 1933. They proved that the necessary and sufficient condition for the transfer function $H(w)$ of a filter to be physically realizable is that

$$\int_{-\infty}^{\infty} \frac{|\log |H(w)||}{1 + w^2} dw < \infty.$$

In particular, the transfer function cannot have too many zeros. Specifically, its zeros on the real axis must necessarily be isolated. Consequently, ideal filters are not physically realizable, and their approximation by realizable filters is one of the most significant and important problems in electrical engineering.

At first, every engineer will be surprised by the Paley–Wiener Theorem's statement because the transfer function's phase does not appear at all (the theorem depends exclusively on its modulus). However, the result is correct because when we restrict ourselves to causal filters, the phase and the modulus of the transfer function are not independent. On the contrary, once one of these functions is known, the other is determined. This fact is a fundamental one, and we owe its discovery to Wiener.

The convolution operators $L(x)(t) = \int_{-\infty}^{\infty} x(s)h(t - s)ds$ always satisfy the property of invariance by translations in time. Indeed, there is a famous theorem — the so-called Schwartz kernel theorem — that shows that under minimal conditions, every wave filter is necessarily a convolution operator. In other words, every filter that has some physical interest is necessarily a convolution operator. Therefore, the convolution operators are also of interest when we change the spaces of input and output signals. For example, to guarantee the stability (and, therefore, its character as a wave filter) when both the inputs and outputs belong to $L^\infty(\mathbb{R})$, it will suffice to require that $h \in L^1(\mathbb{R})$. This property is usually called BIBO (from "Bounded Input, Bounded Output") stability and, in the case that the transfer function $H(w)$ is a rational function, the associated system is BIBO stable if and only if all poles of $H(w)$ have a negative real part.

Wave filters are undoubtedly very useful in engineering. Let us explain one of their most straightforward applications so their importance in daily life becomes clear. Suppose that we want to broadcast several channels simultaneously through a radio antenna so that a receiver is subsequently able to receive any of these channels independently and freely. How do we do this? The idea is very simple. If the signal emitted by the kth channel is the function $x_k(t)$, which we suppose to be a signal with finite bandwidth less than or equal to 20,000 Hertz — because that is the limit imposed by our hearing — then we make our antenna emit the signal $x(t) = \sum_{k=1}^{n} e^{iw_k t} x_k(t)$, so that the signal emitted by each channel is modulated, in the time domain, becoming a signal of the type $e^{iw_k t} x_k(t)$, which is a band-limited signal with the bandwidth of the original signal but with a centered frequency interval on the point w_k. This has the effect that if we take the values such that the intervals

$$[w_k - 20.000, w_k + 20.000]; \quad k = 1, 2, \dots, n$$

are pairwise disjoint, then we can make our receiving device filter the received signal with band pass interval $[w_k - 20.000, w_k + 20.000]$ and then multiply the received signal by $e^{-iw_k t}$, thus obtaining precisely the original signal $x_k(t)$.

This process is called amplitude modulation and it is the method used to create the AM radio. Although the ideal filtering imposed here is not physically realizable, it can be very well approximated by filters that are physically realizable and, furthermore, taking into account that the input signals are already band-limited from the beginning and that we can separate the intervals $[w_k - 20.000, w_k + 20.000]$ at pleasure, this method recovers the emitted signals with complete precision.

In 1929, long before proving his theorem with Paley, Wiener became concerned with another important aspect of wave filter theory: Would it be possible, knowing that a given filter is physically realizable, to build an electrical circuit that simulates it with precision as large as one wishes?

Wiener had an apparently simple idea that — he was sure — would work. It was about using the fortunate fact that Laguerre's functions, which are given by

$$L_n(t) = \sum_{k=0}^{n} 2^{n-k+\frac{1}{2}} \frac{n!}{k![(n-k)!]^2} t^{n-k} e^{-t}, \quad n = 0, 1, 2, \ldots,$$

have the following two important properties:

(a) They form an orthogonal basis of the space $L^2(0, \infty)$, which is the space of the transfer functions of the causal filters that carry signals of finite energy in bounded signals (A direct proof of this last fact can be consulted in the author's paper [AR]).

(b) For each natural number n, the Fourier transform of the function $L_n(t)\chi_{(-\infty,0]}(t)$ is the transfer function of a physically realizable filter "in the metal." In other words, it is a filter for which we know an electrical circuit that simulates it with total precision.

This means that if the transfer function of a filter is physically realizable, then we can decompose it as an (infinite) sum of transfer functions that are realizable in the metal and, therefore, we can approximate it as precisely as we want.

The idea was very good and, at least from a mathematical point of view, totally correct. However, it was necessary to go one step further and build the electrical circuit described by the theory. Obviously, this required an engineer's intervention since Wiener was famous for his lack of manual skills. So, Wiener went to talk to Bush, who was one of the most influential members of the electrical engineering department at MIT, to ask him to look for a student who would like to do a doctoral thesis on this particular problem.

In his autobiography, Wiener would warmly thank Bush for finding not just any student but probably the best possible student for this kind of problem. It was the electrical engineer Y. W. Lee (1904–1989), with whom Wiener would maintain contact for the rest of his life and of whom he would comment as follows:

Yuk Win Lee
(1904–1989).
Courtesy:
M.I.T. Museum.

Lee and I have been scientific colleagues now for about a quarter of a century. From the beginning, his steadiness and judgment have

> furnished exactly the balance wheel I have needed. My first idea
> of an adjustable corrective network would have worked, but at
> the cost of a great wastefulness of parts. It was Lee who saw how
> the same part could be used to perform several simultaneous
> functions and who in that way reduced a great, sprawling piece
> of apparatus into a well-designed, economical network.

At first, Lee found it difficult to understand Wiener's idea, and
his attempts to apply it failed. He was stuck for a year, but when
he finally understood the nitty-gritty, he knew how to put Wiener's
ideas into practice, and they worked so well that no one believed it.
The experiments had, at first instance, failed because Lee had not
realized that to fix the transfer function of a causal filter, it must be
taken into account that its real and imaginary parts cannot be independent, but on the contrary they form a transformed pair through
the so-called Hilbert transform. This was, in fact, the first time that
this transform was applied to an electrical engineering problem, and
precisely because of this, MIT engineers were not at all familiar with
this technique. What's more, Lee's thesis results were so good that he
had severe difficulties defending it. According to one of his students:

> When Lee did his doctoral thesis presentation, there were
> twenty [tenured] professors in the electrical engineering depart-
> ment and they all attended the defense. They all jumped on
> him and pushed him to the max. They were the heavy hitters
> at MIT so Lee, this cultured Chinese, started to retire. Finally
> Wiener, who was still a young man at the time, got up and
> said: "Gentlemen, I suggest that you take this document home
> and study it for yourself. You will find that it is correct." Those
> were his words and Lee finished the thesis exam this way. (...)
> Two weeks later, Lee found a short note in his mailbox: "You
> approved." They didn't have the decency to say, "Hell, what a
> contribution!" That's all he got.

However, this thesis marked a milestone in the history of electronic engineering. In the manuscript, it was shown that being physically realizable and being realizable in the metal are the same thing,
and this was completed by building the physical system that allows
any physically realizable filter to be approximated with the desired
precision. This device was called the "Lee–Wiener machine."

The invention was magnificent. Both Lee and Wiener and the
other MIT engineers were soon aware that the Lee–Wiener machine
could be instrumental in numerous research and technological

advances. This motivated the authors to create a patent in 1935, which would be followed by two more patents between 1935 and 1938, which caused them many headaches. According to Wiener, thanks to Lee's tenacity, they overcame the innumerable legal and bureaucratic pitfalls systematically posed by the Patent Office. Wiener was so traumatized by the labyrinthine process involved in creating a patent that he complained about it several times in his various writings, especially in his book on the process of invention, in which he devoted an entire chapter to his complaints. In fact, Wiener was very unlucky with his patent. Once granted, the Bell Labs showed interest in buying the patent. Wiener and Lee were seduced by the possibility that such an important company might use their invention. They thought this would serve to publicize their finding, which led them to agree to a low price (about $22,000). They didn't expect that the company would put their patent in a drawer for the next seventeen years. Bell Labs' sole interest in Lee–Wiener's invention was to head off stiff competition from other telephone companies. Wiener was devastated and, from that point on, developed a growing resentment towards the company.

After completing his doctoral thesis, Lee had no luck in his job search in the United States. The university offered him nothing, which forced him to work as a development engineer for the United Research Corporation (later renamed Warner Brothers) on Long Island. Wiener sadly comments in his autobiography that despite his best efforts, he was unable to find a good job for his student, among other things because:

> At that time, the Asian-born engineer already existed in the United States, but he was a much rarer bird than he is today. The resistance I had to face was far greater than what I could overcome, and Lee had to return to China in search of work, first in industry and later in academia.

So, Lee had to go back to his country. Indeed, his life went through a very complicated period worthy of a fiction film from his doctoral thesis's defense until the end of the Second World War. In China, he first obtained a job as an electrical engineer for the China Electricity Company in Shanghai and, after marrying Elisabeth — a beautiful Canadian citizen he had met in the United States — in 1933, in 1934, he secured a teaching position at the National Tsing Hua University in Beijing. Three years later, while visiting his parents in

Hangchow, Japan invaded northern China, and the Lees were trapped in Shanghai, unable to return home. They lost all their belongings and were forced to survive in Shanghai, which was made possible by setting up a small antique shop. As he had completely given up hope of returning to his former position at Tsing Hua University, Lee wrote to Wiener asking for his help. This time Wiener got MIT to offer him an assistant professor position, but just when the Lees were ready to go to the United States — they had already bought tickets to travel by boat from Hong Kong — the bombing of Pearl Harbor took place. This, coupled with administrative negligence, forced Lee and his wife to stay in China under the Japanese mandate until the end of the war, experiencing real misery. Finally, in 1946, the Lees came to MIT and would stay there forever.

Lee had a brilliant academic career and, among other things, led an important research group devoted to the study of the statistical theory of communication. In particular, he managed to disseminate among engineers with great efficiency the research carried out by Wiener in this new branch of mathematics. Some of today's leading scientists in the area are direct descendants of Lee or his students. For example, A. V. Oppenheim (born in 1937), who is the author of some of the best books on signal theory for telecommunications engineering courses, is Lee and Wiener's scientific grandson (and, therefore, great-grandson of Wiener).

When Lee was employed as a professor at the National Tsing Hua University, he had time to invite Wiener to stay there as a visiting professor during the academic year 1935–1936. Wiener was delighted with the idea of "Travel to the East" and was also confident that he could advance his investigations with Lee. Furthermore, he was expected to deliver a series of lectures on generalized harmonic analysis and his work with Paley on the complex domain Fourier transform and its applications. Wiener decided to travel with the whole family.

As for his work with Lee, it was a dismal failure. As he would recall in 1956:

> What Lee and I had really tried to do was to follow in the footsteps of Bush in making an analogy-computing machine, but to gear it to the high speed of electrical circuits instead of to the much lower one of mechanical shafts and integrators. The principle was sound enough, and in fact has been followed out by other people later. What was lacking in our work was a

> thorough understanding of the problems of designing an apparatus in which part of the output motion is fed back again to the beginning of the process as a new input. This sort of apparatus we shall know here and later as a feedback mechanism.
>
> (...) What I should have done was to attack the problem from the beginning and develop on my own initiative a fairly comprehensive theory of feedback mechanisms. I did not do this at the time, and failure was the consequence.

Apparently, it was then that Wiener first became interested in feedback systems. The study of such systems was undoubtedly one of the fundamental axes around which the new discipline that he founded a decade after his visit to China and which he called cybernetics would revolve.

Feedback Systems

A system's feedback consists of using the system's output to modify its input according to a specific objective. This type of operation is critical in control theory. The idea is that the system itself has a particular automatic criterion to modify the inputs in order to keep the outputs within a set of margins in advance. So, let's consider a linear system $L\colon X \to Y$ and take a new linear system $K\colon Y \to X$ (which we will design according to some specific objective that we have set ourselves), then we modify the input of the system L so that the new input is $e = x - Ky$. This results in the pair of equations

$$\begin{cases} y = Le \\ e = x - Ky \end{cases},$$

which, viewed simultaneously, describe what we call a linear feedback system. What is the effect of the above equations on each input $x \in X$? To see this, we get the input x from the second equation, which leads to $x = e + Ky$, and then substitute $y = Le$ to get that

$$(I + KL)e = x. \tag{1}$$

This functional equation, which is also sufficient for the complete description of the feedback system, tells us that if the original input function of the system is x, then after it is modified, the new input is the solution (if it exists) e of the functional equation (1).

Usually, the theory of feedback systems is explained for causal analog systems that transform one-dimensional (time-dependent) signals into one-dimensional signals. Hence, the elements of the function spaces X, Y are usually functions defined on $[0, \infty)$. The system described by the functional equation (1) is said to be regular (with respect to the spaces X, Y) if the linear operators $L : X \to Y$, $K : Y \to X$ are both bounded, and for each value of $T > 0$ and each input $x \in X$, the equation

$$(I + KL)e_T = x_T, \tag{2}$$

where

$$x_T(t) = \begin{cases} x(t) & (t \leq T) \\ 0 & (t > T) \end{cases},$$

has a unique solution $e_T \in X$ and, moreover, for each $t > T$ we have that $e_T(t) = 0$.

The system is called stable if there is a constant $M > 0$ such that for every input $x \in X$, the only solution $e \in X$ to (1) satisfies the inequality $\|e\|_X \leq M \|x\|_X$. It can be shown that the stability of the regular feedback system (1) is equivalent to the fact that the operator $I + KL$ admits a continuous inverse (or, if one prefers to express it symbolically, that $-1 \notin \sigma(KL)$, where $\sigma(KL)$ denotes the spectrum of the linear operator $KL : X \to X$).

What does the above condition mean when we are considering wave filters? If the transfer function of the filter L is $H(w)$ and the transfer function of K is $G(w)$, then the functional equation (1), in conjunction to $y = Le$, tells us that the transfer function of the feedback system is given by

$$R(w) = \frac{H(w)}{1 + H(w)G(w)}.$$

Therefore, when $R(w)$ is a rational function, the system's BIBO stability will be guaranteed when all the poles have a negative real part. Of course, there are general criteria for BIBO stability (and other types of stability) that don't require that $R(w)$ be a rational function. These criteria use complex variable arguments and topological arguments, and Wiener, who knew them, used them to develop cybernetics from 1948 on.

However, his stay in China would bring him unexpected fruit. While reviewing the material that he would use to teach his classes, Wiener became interested in the theory of quasi-analytic functions. Despite not admitting a power series expansion, these functions share certain stiffness properties with the complex variable analytic functions. The most important of them is that once they are known on a non-empty interval, independent of how small it may be, they are determined on all its domain of definition. Then, he had a stroke of luck: J. Hadamard (1865–1963), the great French mathematician, visited the university where Wiener was invited. As a result of the meeting, Wiener had the opportunity to speak at length with him. In this way, he got Hadamard to put him in touch with his student Szolem Mandelbrojt (1899–1983) — who was a specialist in quasi-analytic functions — and arrange things for a meeting just before the ICM of 1936, which would take place in Oslo and which Wiener planned to attend.

N. Wiener at the 1936 ICM Gala Dinner, Oslo (Wiener at far right).

Wiener and Mandelbrojt, who met in Paris, quickly became friends and over the years published several interesting joint papers.

Sometime later, in 1952, Benoit Mandelbrojt (1924–2010) — now known as the great promoter of fractal geometry and its many applications — who was Szolem's nephew, would travel to MIT with the idea of working with Wiener whom he had admired since childhood under the influence of his uncle. But he found Wiener in a state of deep isolation and depression and they could not do anything together.

Leo Wiener had retired as a professor several years earlier, although he maintained his research activity, attending the Harvard library daily. Shortly after his retirement, he had suffered a relatively severe accident (he was run over, breaking his femur), and since then had lost much of his vitality, but he was still active. However, on his return from China, Wiener found that his father had deteriorated.

> He needed hospital care, but this time there was much less sign of recovery than there had been on the previous occasion. He lapsed into an agitated state of depression, in which his mind was often confused. Yet he was fully aware that he was confused and that he was losing his grip on life (. . .).
>
> I visited Father often, and took him out from time to time for trips in my car. However, he was on the way down, and it was scarcely even desirable that this half-life, which was all that was left to him, should be drawn out indefinitely. Finally, in the first year of the war, he died calmly and peacefully in his sleep.

Tauberian Theorems, Number Theory, and the Bôcher Prize

As we already know, in 1926, Wiener was involved in the construction of his generalized harmonic analysis, a subject in which he had become "stuck" because of a particular technical result on the averages of positive functions that he was not able to prove. As he himself would recall in his autobiography,

> There were places in my theory of generalized harmonic analysis which I was nearly but not quite able to bring to a definite close. I needed certain theorems, and I found myself proving similar but not identical ones.

Those days were also tough for Wiener because they coincided with the season in which he felt humiliated by the treatment he received in Göttingen. But then, he was lucky to meet A. E. Ingham (1900–1967), a Cambridge-era friend of his who was an expert in number theory, who, once aware of the technical difficulties Wiener was facing, reoriented his thought. He made it clear that the problems Wiener was facing were of the same type as a series of mathematical problems on which Wiener's dear friend and teacher Hardy, in the company of his colleague Littlewood, had spent considerable time. They had proved an important battery of identities using what they called the Tauberian theorems method.

What kind of problems were these? A classical result on the convergence of power series is the so-called Abel's Theorem, according to which, if

$$\sum_{n=0}^{\infty} a_n = A \tag{1}$$

is a convergent series of numbers (real or complex), then

$$\lim_{x \to 1^-} \sum_{n=0}^{\infty} a_n x^n = A. \tag{2}$$

A power series that satisfies this last equation is said to converge in the Abel sense to the value A at the point $x = 1$. In 1897, A. Tauber (1866–1942) proved that if (2) is verified and we add the hypothesis that

$$\lim_{n \to \infty} n a_n = 0, \tag{3}$$

then (1) is also satisfied.

It soon became clear that the condition imposed by Tauber was excessively restrictive, but the search for the optimal conditions to impose on the coefficients a_n that would guarantee, from Abel's convergence of the power series, the convergence of the coefficients, alone, was a subject of enormous technical difficulty. Before Wiener came on the scene, the best results were due to Hardy and Littlewood's joint effort. In particular, they had managed to reduce the hypothesis (3) to the less restrictive $\sup_{n>0} n |a_n| = K < \infty$ and, later, they reduced the latter hypothesis to the most general condition: $n a_n > -K$ $(n = 0, 1, 2, \ldots)$ for a certain constant K.

These results, intended to demonstrate the convergence of a certain numerical series (or, also, an integral) as a consequence of the validity of certain constraints on the coefficients of the series (or the function to be integrated), are called Tauberian theorems. Hardy and Littlewood had studied the asymptotic behavior (for $y \to 0^+$) of numerous integrals of the type

$$\int_0^\infty \varphi(x/y)\phi(x)dx$$

and Wiener realized that if we change the variables in such a way that $x = e^{-s}$, $y = e^{-t}$ and introduce the new functions, $f(u) = \varphi(e^u)$, $g(u) = e^{-u}\phi(e^{-u})$, then

$$\int_0^\infty \varphi(x/y)\phi(x)dx = \int_\infty^{-\infty} \varphi(e^{-s}/e^{-t})\phi(e^{-s})(-1)e^{-s}ds$$

$$= \int_{-\infty}^\infty \varphi(e^{t-s})\phi(e^{-s})e^{-s}ds = \int_{-\infty}^\infty f(t-s)g(s)ds,$$

thus transforming the original integral into the convolution

$$(g * f)(t) = \int_{-\infty}^\infty f(t-s)g(s)ds.$$

This leads us to study a limit of the type $\lim_{t \to +\infty}(g * f)(t)$ as an asymptotic problem. This suited Wiener very well, as he had managed to keep the accounts on his own ground, Fourier analysis. Thus, in Wiener's formulation, the problem to be studied was the following one: Once the functions $f(t), g(t)$ are fixed, when can we guarantee the existence of the limit $\lim_{t \to +\infty}(g*f)(t)$? Or, in other words, given the wave filter $L(f) = g*f$, what inputs f give rise to an output $L(f)$ with the property of having a precise limit behavior for $t \to +\infty$?

Wiener demonstrated the following general result:

Theorem (Wiener's Great Tauberian Theorem). Suppose that $f \in L^\infty(\mathbb{R})$ and $g \in L^1(\mathbb{R})$. If there is a new function $g_0 \in L^1(\mathbb{R})$ such that its Fourier transform has no zeroes on the real line and, furthermore, the relation

$$\lim_{t \to +\infty}(g_0 * f)(t) = A \cdot \int_{-\infty}^{+\infty} g_0(s)ds, \qquad (1)$$

holds, then the identity

$$\lim_{t\to+\infty} (g * f)(t) = A \cdot \int_{-\infty}^{+\infty} g(s)ds, \tag{2}$$

is also satisfied (with the same constant A).

Conversely, if the function $g_0 \in L^1(\mathbb{R})$ has the property that $\int_{-\infty}^{+\infty} g_0(s)ds \neq 0$ and for every pair of functions $f \in L^\infty(\mathbb{R})$ and $g \in L^1(\mathbb{R})$ the implication (1) $\Rightarrow$ (2) holds, then the Fourier transform of g_0 has no zeroes on the real line.

He then proved a valid version of the theorem for integrals of the type $(g * \eta)(t) = \int_{-\infty}^\infty g(t - s)\eta(ds)$, where we have replaced the function f by a measure η defined on the real line. It is precisely the use of general measures that makes it possible to establish a bridge between problems on estimating improper integrals and problems on the convergence of numerical series.

Obviously, the proof of Wiener's Great Tauberian theorem is not trivial. We would like to comment that it was based on a wide range of technical lemmas, one of which is especially remarkable not only for its beauty and the simplicity of its statement but also because, by itself, it had an important impact among analysts, giving rise to one of the main motivations for the creation of the theory of Banach algebras by the Russian mathematician I. Gelfand (1913–2009) in 1941. It is the following technical result:

Lemma (Wiener's Little Tauberian Theorem). Suppose that $f(t)$ is a continuous and 2π-periodic function such that $f(t) \neq 0$ for every real number t, and that $\{c_k(f)\}_{k=-\infty}^{k=\infty}$ denotes the sequence of the Fourier coefficients of $f(t)$. Then,

$$\sum_{k=-\infty}^{\infty} |c_k(f)| < \infty \Leftrightarrow \sum_{k=-\infty}^{\infty} |c_k(1/f)| < \infty.$$

He approached the proof of this lemma in a very "head-on" way: He calculated the Fourier series expansion of $1/f$ from the Fourier series of f and made several estimations that led him — going through some more or less delicate tricks, such as multiplication by a trapezoidal function or the use of partitions of unity — very elegantly to the desired result.

His proof did not stray even a millimeter from the idea that he was working with functions. Perhaps this prevented him from seeing the problem from a more abstract point of view — as Gelfand would later do — as a question about "closed invertibility" in Banach algebras. If he had done this, perhaps he would have founded this new branch of functional analysis. In any case, there is still another point of view that makes us see the previous lemma as something absolutely natural. Concretely, it is well known that there is a strong link between a function's smoothness and the speed with which the Fourier coefficients $c_k(f)$ of a function f converge to zero for $k \to \pm\infty$. How rapidly the $c_k(f)$ converge is important in Electrical Engineering. Thus, to say that $\sum_{k=-\infty}^{\infty} |c_k(f)| < \infty$ can be literally interpreted as a concept of smoothness for f. So, Wiener's lemma tells us that if we can compute $g = 1/f$, this new function will be just as smooth as the function f, which is after all completely natural.

In a joint work with U. Luther, the author of this biography had the privilege of proving (see [AL]) that for all the concepts of smoothness existing up to now, the functions $f(t)$ and $1/f(t)$ always have the same degree of smoothness.

Wiener's Tauberian theorems' importance is not limited to the indisputable fact that they were original and technically complex. What is truly crucial and what led to the international mathematical community's astonishment, was that these results, so simple in statement, not only served to rescue as particular cases all the Tauberian theorems existing to date but also proved some theorems that until then had resisted a proof.

An important example of this was a Tauberian theorem relative to the Lambert series from which a method was known to prove the prime number theorem, one of the most profound results of modern mathematics. Other examples included a battery of results on the convergence of Fourier series, and some theorems relative to the closure properties of certain trigonometric systems. In particular,

Wiener achieved his initial goal of putting his general harmonic analysis on solid ground.

As soon as he learned of Wiener's successes, his friend Tamarkin, then a Professor at Brown University, persistently encouraged him to write a couple of monographs on the new (and now firmly established) Fourier analysis and on the Tauberian theorems. Wiener listened to him, and both monographs eventually appeared in the form of long articles. The first of them contained his results in generalized harmonic analysis. It was published in 1930 in *Acta Mathematica*. The second, which appeared in the *Annals of Mathematics* in 1932, was devoted to Tauberian theorems. It was the publication of these two memoirs that finally pushed Wiener into the Olympus of first-line mathematicians. What he had not achieved with his previous contributions concerning the Brownian motion, operational calculus, or potential theory, namely, recognition from his North American colleagues, was obtained this time without a hint of doubt. This recognition became clear in 1933 when the American Mathematical Society (AMS) decided to award him the Bôcher Prize. In addition, in April 1934, he was named a Fellow of the United States National Academy of Sciences, another important public recognition, and shortly after that, he was appointed Vice President of the AMS, his enormous dislike for administrative matters being the only reason for his not reaching the presidency. In April 1933, just a few months after Hitler's appointment as chancellor, the German government published a Law for the Restitution of the Civil Service whose sole purpose was to expel Jews from the civil service (anyone who had at least one Jewish grandparent was considered a Jew). This, obviously, would cause a stampede of university professors looking for a position outside of Germany, with the United States being the country that received the most. Wiener, who since 1932 occupied a chair at MIT and was a respected mathematician, did not hesitate to take all kinds of steps to relocate those mathematicians who asked him for help. In particular, it is well known that he actively participated in the relocation of O. Szász (1884–1952), H. Rademacher (1892–1969), G. Pólya (1887–1985), G. Szegö (1895–1985), and K. Menger. 1933 was also an important year for Wiener because precisely then he met whom we could consider his lifelong best friend and one of his sources of inspiration for the creation 15 years later of cybernetics, Arturo Rosenblueth. We will talk about the work they did together and their relationship in Chapter 6. Although Wiener was admired in life (and after death) by engineers for his contributions to wave filter theory, prediction theory, cybernetics, Brownian motion, etc., mathematicians worldwide know him mainly by his Tauberian theorems. That single contribution would have been enough to grant him a privileged place in the history of mathematical analysis.

Chapter 5

Mathematics for War

A Digital Computer Project

Vannevar Bush, the MIT engineering department professor with whom Wiener collaborated in the 1920s and who later became vice president and dean of the Engineering Faculty at MIT (1932–1938), accepted the appointment as president of Washington's Carnegie Institution in 1939. This institution was responsible, among other things, for granting large amounts of money for research in the United States. This appointment and the acceptance that same year of the position of director of the National Committee for the Aeronautical Council obviously allowed him to influence the direction that investigations would take in his country from 1939 on. His decision, undoubtedly, was that the country needed its scientists and engineers to carry out military-oriented projects. Bush was concerned about the limited scientific collaboration that had always existed between the military and civilians. He thought that, given the circumstances and, in particular, considering that the United States would probably end up participating in the fierce conflicts that were ravaging Europe at that time, the fluid cooperation of all parties was essential, as well as a broad mobilization of civilian personnel towards defense-related projects. Thus, he proposed creating a national commission for research in defense matters. For this, he requested to meet with President Roosevelt, to whom he presented on June 12, 1940, a single sheet of paper with his reasoned proposal. Roosevelt did not need even 10 minutes to approve the project. The NDRC (National Defense Research Committee) was officially created by order of the

National Defense Council on June 27, just a few days later, and with Bush as president. At a certain point in the early 1940s, Bush had working under him — as chairman of the Office of Scientific Research and Development, which would be created in 1941 and would absorb the NDRC, in addition to taking care of the leadership of the Manhattan Project until 1943 — two-third of the physicists involved in research in the United States.

Bush soon sent a circular to the different departments of engineering, physics, mathematics, etc., with a form requesting proposals for action to carry out research projects related to national security. Wiener was one of the first mathematicians to answer. The first thing he did was advise Bush on how he thought the different teams should be organized. Specifically, Wiener believed that research projects must necessarily be interdisciplinary and voluntary. On the other hand, they should be made up of small groups of people and not be restricted by an excessive hierarchy. Finally, the different teams would meet from time to time to carry out joint attacks on the diverse problems they faced.

Next, Wiener began to think about some concrete action proposals. Bush had become a top-of-the-line engineer mainly because of his ability to build different analog machines and, in a very particular way, because of the construction, in the late 1920s, of a Differential Analyzer, which was an artifact thought to approach the solution of ordinary differential equations. In this machine, the different numerical quantities involved in the equations were represented by introducing specific physical quantities (such as voltages, resistances, etc.) whose physical interaction is precisely described by the differential equation to be solved. The mathematical relationship between the differential equation's different variables was represented by "analogy" with these physical quantities.

The first differential analyzer that Bush built was used for the calculation of integrals of the type $y(t) = \int_0^t f(s)g(s)ds$, where the graphs of the functions $f(t), g(t)$ are the data and the function $y(t)$ is the response of the system to these data. By 1927, Wiener had already helped Bush in his work, devising a different mechanism that allowed the calculation of integrals in which the parameter t intervenes within the integrand, such as the integrals of the form $I(t) = \int_0^t f(t-s)g(s)ds$ which are so important in harmonic analysis. The device, which unlike Bush's was optical and not mechanical in nature, received the name Wiener's Cinema Integraph. It was built by K. E. Gould in 1929 and led to the completion of numerous doctoral theses at MIT, all aimed at including improvements either in the execution time of machine tasks or in expanding the range of possible applications. By 1940, they had already succeeded in making the apparatus compute quite efficiently integrals of the type $F(t) = \int_0^t f(t,s)g(s)ds$.

Although Bush's machine was considered a complete success, it had an apparent problem: There was no way to make it work to solve partial differential equations. And precisely these equations were of vital importance for mathematical physics and, in particular, for problems related to war. Wiener, who after the conception of his Cinema Integraph had worked with Lee on the design of several analog machines, and in particular perfectly knew the execution times of each of the basic components, realized that one of the main difficulties in building a specific-purpose analog machine that would solve a PDE lies precisely in the analog character of the machine. Taking all this into account, he considered that an important project, from a military point of view, was to construct a digital device that would solve these equations.

The main reason why an analog machine could not function properly is that electromechanical devices (such as spinning wheels, switches, and punch cards) must move and are subject to significant inertia and friction. This considerably slows down the execution time. And, in the case of partial differential equations, any delay in the execution of a simple task such as multiplication or the addition of two quantities is very detrimental, because when dealing with problems in two or more dimensions, each of these operations must be carried out on a whole mesh of points, and that implies a lot of data.

However, Wiener had an idea to tackle the solution to this problem. Specifically, he was convinced that the "scanning" method used to read and transmit television images could very well be used to read the values of a defined function on a planar region. If this were achieved, then likely most of the process of calculating numerical solutions for the different PDEs that appear in mathematical physics and engineering could be accelerated.

Of course, Wiener had not only thought about how to improve computing speed. As the machine was to be a digital one, in each operation, all the significant digits necessary for the quantities involved could be calculated with a precision as great as desired. This would be a clear advantage compared to the analog machines, where all numbers are based on measurements of physical magnitudes that could only be an approximate reflection of the true values to be computed. He had also decided that the machine to be built should perform all its computations in the binary system instead of the usual decimal system that everyone was used to. The binary system has the property that the addition and multiplication tables are very simple, and therefore are easy to implement. In particular, they can be run very quickly with vacuum tubes, one of Wiener's preferred devices. Furthermore, in terms of controlling calculation errors, this system is much more efficient than any other. Finally, Wiener gave a theoretical description of the computer he was thinking about.

Wiener presented his work project in September 1940 in the form of a 12-page report entitled "Memorandum on the Mechanical Solution of Partial Differential Equations" and a four-page letter outlining the project's motivations which were addressed directly to professor Bush. In his description of the machine whose construction he proposed, he said the following:

> The projected machine will solve boundary value problems in the field of partial differential equations. In particular, it will determine the equipotential lines of flow about an airfoil section given by determining about 200 points on its profile, to an accuracy of one part in a thousand, in from three to four hours. It will also solve three-dimensional potential problems, problems of the theory of elasticity, etc. It is not confined to linear problems and may be used in direct attacks on hydrodynamics. It will also solve the problem of determining the natural modes of vibration of a linear system.

However, it seems that Bush did not find this project interesting.

> I made a report of my suggestion to Vannevar Bush, but I did not get a very favorable reception. Bush recognized that there were possibilities in my idea, but he considered them too far in

the future to have any relevance to World War II. He encouraged me to think of these ideas after the war, and meanwhile to devote my attention to things of more immediate practical use.

Later on I found that he had no very high opinion of the apparatus I had suggested, especially because I was not an engineer and had never put any two parts of it together. His estimate of any work which did not reach the level of actual construction was extremely low. The only satisfaction I can now get is that I was right something like ten years before the techniques to prove my ideas were developed.

Antiaircraft Batteries, Filters, and Prediction Theory

Since the project to build a digital computer was not considered valid for the present war interests, Wiener set out to think of other possible ways of collaboration. The first option he thought of was the possibility of studying the problem of encoding and decoding messages. However, he ended up opting for a very different question: the study of possible improvements in antiaircraft batteries' operation. The problem is quite complex because it involves predicting the future trajectory of an aircraft in combat. Indeed, since these planes are so fast, their speed is a substantial part of the projectile's speed with which we intend to shoot it down. Therefore, at the firing time, we must not aim at the position that, according to the radars, the aircraft occupies precisely at that moment. On the contrary, we must guess what position the aircraft will occupy in the relatively near future. If we do not have some additional information, it is impossible to predict the future values of a signal of which we only know its past and, probably, its current value. Only if we know something more than the signal's past is it perhaps reasonable to venture into a description of the future. To give some intuitive examples, if $x(t)$ is a line, a parabola, or any other polynomial, it is very easy to determine the behavior (past and future) of $x(t)$ from a finite number of its values: Just take $n + 1$ points, if the degree of the polynomial is n. Another fundamental example is the case where the signal we are studying is analytical. In this case, knowing it over a finite interval or simply on a sequence of points that converges to a point inside the signal's holomorphy domain allows us to identify it in its entirety (how to do this is another question). However, it is clear that we cannot assume that the trajectory described by a fighter plane is an analytical function. It is quite natural to think that it may have some peaks and, therefore, lose the property of being differentiable at certain points. On the other hand, we cannot affirm that a trained pilot can carry out any type of trajectory, since if one does not maneuver with some care, it is very easy to destabilize the aircraft. So, although the planes' trajectories are not smooth (and therefore predictable), they are not arbitrary functions either. It is precisely this aspect that allows us to believe that an adequate mathematical theory could exist for the problem of predicting airplane trajectories.

The first person with whom Wiener spoke about his project to build a device that would improve the action of antiaircraft batteries was Professor S. H. Caldwell (1904–1960), who was in charge of coordinating all activities aimed at what were called "war problems" and which used the Bush differential analyzer that they had set up at MIT. Caldwell imposed on Wiener an absolute silence on his ideas on this subject, which automatically became a state secret, and simultaneously began to work with him to carry out some first experiments that would shed light on the usefulness of Wiener's ideas. They experimented with the differential analyzer for three weeks and finally came to the conclusion that there was indeed a chance of success, although the work was still in a rather initial and hypothetical state. Caldwell and Wiener sent a brief job proposal to the NDRC, which shortly before Christmas 1940 accepted the project, granting the small amount of money that was requested for its realization (2,325 dollars). This is how Wiener found himself at the helm of DIC project 5980 on antiaircraft batteries.

By chance, during that same period, Julian Bigelow (1913–2003), an MIT-trained engineer who had worked for IBM for several years, stopped by the MIT engineering department to present his credentials, and perhaps to get some kind of job that would help him request a postponement for his eventual conscription into the army, given that a general "call-up" was expected on the occasion of the entry of the United States into the war. To his surprise, K. Wildes, the department director, not only agreed with him on the adequacy of his request for the postponement but actually promoted it himself. In Bigelow's words, Wildes appears to have told him:

> You cannot go to the army. We need you to stay on here and
> work with Wiener, see if you can help him finish what he's
> trying to do, or what he's already doing. No one here has a clue
> what he's been talking about lately.

In this way, after an interview with Warren Weaver (1894–1978), then director of the Division of Sciences of the Rockefeller Institute and in charge of section D-2 on fire control in the NDRC, Wiener was assigned an engineer to help him with his project. Apparently, Wiener was lucky again (as he had been with Lee), and Bigelow was the right person for the job at hand.

The starting point for their discussions was the study of classic parabolic shooting problems and the application of these to clay pigeon shooting problems and similar issues. In these cases, of course, the projectile velocity is much higher than that of the moving target. Therefore, the need for prediction is much less. They then made several visits to military bases off the coasts of Virginia and North Carolina. These excursions were aimed at conducting a field study. They took measurements and checked, on the ground, what kind of trajectories a fighter plane could follow. At this point, they realized the enormous difficulty of the problem they faced.

But Wiener had an idea that apparently would work. He assumed that enemy aircraft trajectories are described by signals of the class S for which he had created the generalized harmonic analysis. So, if $f(t)$ represents the plane's position at the instant of time t, and we consider the approximation of the value $f(t + \alpha)$ by using a causal wave filter, this must be an operator of the form

$$L(f)(t) = \int_0^\infty f(t - s)dK(s)$$

for a certain function of bounded variation $K(s)$. Therefore, we can propose as a measure of goodness for the approximation that we are making, the quantity

$$E(K) = \lim_{T \to \infty} \frac{1}{2T} \int_{-T}^{T} \left| f(t + \alpha) - \int_0^\infty f(t - s)dK(s) \right|^2 dt,$$

which leads us to the search for a function $K(s)$ that minimizes this value. Now, it is not difficult to verify that the limit can be expressed as

$$E(K) = \varphi(0) - 2Re\left(\int_0^\infty \varphi(\alpha + s)d\overline{K(s)}\right) + \int_0^\infty \int_0^\infty \varphi(s - t)d\overline{K(t)}d\overline{K(s)}.$$

(It is enough to take into account the equality $|(z + w)|^2 = |z|^2 + |w|^2 + 2Re(z \cdot \overline{w})$ and use the definition of the covariance of the signal $f(t)$).

Thus, we must find the function $K(t)$ that minimizes the value given by this expression. Here, it is important to observe that we have managed to make the function $f(t)$ disappear, so that the same function $K(t)$ will serve to solve the prediction problem for all functions $f(t)$ whose covariance is given by $\varphi(t)$.

The next step to take is, evidently, to solve the optimization problem. To do this, Wiener uses a standard technique in the calculus of variations and concludes that the function $K(t)$ solves the problem if and only if it is a solution of the integral equation given by

$$\varphi(t + \alpha) = \int_0^\infty \varphi(t - s)dK(s), \text{ for } t \geq 0,$$

where $K(t)$ is the unknown function and $\varphi(t)$ is known.

These types of equations had previously appeared in Wiener's work and therefore he knew how to solve them. Specifically, at the beginning of the 1930s, Wiener had worked hand in hand with the Austrian mathematician and astrophysicist E. Hopf (1902–1983), who at that time was visiting Harvard to study ergodic theory under Birkhoff. They worked on a problem in radiation equilibrium in stellar physics, and this work had led them to study this type of integral equation. In his autobiography, Wiener describes the problem with great simplicity:

> Inside a star there is a region where electrons and atomic nuclei coexist with light quanta, the material of which radiation is made. Outside the star we have radiation alone, or at least radiation accompanied by a much more diluted form of matter. The various types of particles which form light and matter exist

in a sort of balance with one another, which changes abruptly when we pass beyond the surface of the star. It is easy to set up the equations for this equilibrium, but it is not easy to find a general method for the solution of these equations.

The equations for radiation equilibrium in the stars belong to a type now known by Eberhard Hopf's name and mine. They are closely related to other equations which arise when two different physical regimes are joined across a sharp edge or a boundary, as for example in the atomic bomb, which is essentially the model of a star in which the surface of the bomb marks the change between an inner regime and an outer regime; and, accordingly, various important problems concerning the bomb receive their natural expression in Hopf–Wiener equations. The question of the bursting size of the bomb turns out to be one of these.

From my point of view, the most striking use of Hopf–Wiener equations is to be found where the boundary between the two regimes is in time and not in space. One regime represents the state of the world up to a given time and the other regime the state after that time. This is the precisely appropriate tool for certain aspects of the theory of prediction, in which a knowledge of the past is used to determine the future. There are however many more general problems of instrumentation which can be solved by the same technique operating in time. Among these is the wave-filter problem, which consists in taking a message which has been corrupted by a simultaneous noise and reconstructing the pure message to the best of our ability. Both prediction problems and filtering problems were of importance in the last war and remain of importance in the new technology which has followed it.

Indeed, Wiener would concern himself not only with the problem of prediction in the simplified form that we have discussed above but he also wanted to take into account a more realistic situation. Concretely, he also dealt with the case where the radar does not accurately reflect the aircraft's true position since some noise could very well contaminate the signal emitted by it.

In that case, assuming that the function $f(t)$ gives the signal detected by the radar, he decomposes it as $f(t) = f_1(t) + f_2(t)$, where $f_1(t)$ is the true position of the target and $f_2(t)$ is the noise created by the measuring devices. He assumes that we know the covariance function of $f(t)$, $\varphi(t) = \lim_{T \to \infty} \frac{1}{2T} \int_{-T}^{T} f(s+t)\overline{f(s)}ds$, the covariance function $\varphi_{22}(t)$ of $f_2(t)$, and the cross covariance $\varphi_{12}(t)$ of $f_1(t)$

and $f_2(t)$. From these data, the cross covariance function of $f_1(t)$ and $f(t)$ can be found, and is given by $\varphi_1(t) = \varphi(t) + \varphi_{22}(t) - \overline{\varphi_{12}(t)}$. With this information in hand, our problem, which is the minimization of the error

$$E(K) = \lim_{T \to \infty} \frac{1}{2T} \int_{-T}^{T} \left| f_1(t + \alpha) - \int_{0}^{\infty} f(t - s)dK(s) \right|^2 dt,$$

is translated — using the same technique that we applied to the case in which there is no noise — to the search for a solution $K(s)$ of the equation

$$\varphi_1(t + \alpha) = \int_{0}^{\infty} \varphi(t - s)dK(s), \text{ for } t \geq 0.$$

Wiener knew how to extend the techniques that he had developed a few years earlier with Hopf to attack this new equation, solving the general problem of prediction with great elegance. In particular, he found a closed expression for the prediction filter and computed the error committed.

E. Hopf (1902–1983). An Austrian mathematician, he made significant contributions to topology and ergodic theory. He received his doctorate in mathematics in 1926 from the University of Berlin, and in 1929 he got his habilitation with a memoire on mathematical astronomy from the same university. He later secured a Rockefeller Foundation scholarship to study under Birkhoff at Harvard. It was probably there that he contacted Wiener, who supported him to be admitted to the MIT department of mathematics in 1931, where he remained until 1936 when he accepted a professorship at the University of Leipzig.

Hopf's march to Germany precisely at that time, when national socialism was in vogue, was widely criticized. Wiener, however, defended his friend because in those years, salaries in the United States were very low and the prestige of a chair in Germany was not comparable to anything you could find then in the United States. Furthermore, according to Wiener, it was better for such a position to be filled by someone who was not particularly favorable to the Hitler's regime as was true in the case of Hopf. Interestingly, the renowned German physicist Heisenberg decided to stay in Germany in 1933, when Jewish teachers' expulsions began, after consulting M. Planck (1858–1947), moved by a similar argument. Hopf had already obtained some important results on ergodic theory in 1930, and during his years at MIT, continued working on the subject. He also collaborated with Wiener in the study of certain integral equations that quickly received the name Wiener–Hopf and, as we have seen, were used by Wiener in his later theory of prediction. Once in Germany, Hopf published an important monograph on ergodic theory, to which Wiener would repeatedly refer in his writings. Apart from these important achievements, Hopf also made important contributions in areas as diverse as quantum mechanics, topology, differential equations, differential geometry, and statistics. One of his most significant achievements was the solution to the well-known Hurewicz Problem in 1947.

Eventually, he returned to the United States (after a brief tour to several German universities), where he acquired American citizenship in 1949. His last years as a professor were spent at Indiana University, where he retired in 1972.

Wiener and Bigelow weren't the only scientists who embarked on a secret project studying antiaircraft batteries. Simultaneously, under the auspices of Bush, the NDRC had already awarded in 1940 a project with similar characteristics (but for a substantially higher amount) to a mixed group of physicists and engineers from Bell Laboratories and MIT. This project's objective was to study all kinds of applications of the new radar technology that could be relevant for national defense and obviously this implied, among other things, the improvement of antiaircraft batteries. To hide the word "Radar," the team working on this other project received the curious name of "Radiation Laboratory" (at that time scientists were not yet aware of how dangerous nuclear physics research could be, which they considered harmless). The Radiation Laboratory depended directly on the NDRC. Still, very soon, they reached an agreement with Karl Compton so that MIT would take care of all the administrative issues related to the Laboratory, in addition to hosting its members. Even so, and even though Wiener and Bigelow worked in the nearby building D-2, where they had set up their mini-laboratory in an empty classroom, they were not aware (nor were they informed) of the existence and objectives of the Radiation Laboratory. They even met at the Bell Laboratories facilities, when they had already been working for five months, with some members of the other team. The meeting was prepared by Weaver, who told Wiener and Bigelow that the other team was also working on the problem of antiaircraft batteries. However, the meeting was a failure. According to Bigelow, they felt used:

> They told us very little [about what they were doing]. They wanted us just to tell them what Wiener's ideas were and how to realize them.

The Radiation Laboratory would soon become visible to Wiener and Bigelow. In a short time, it grew considerably and, in fact, forced the MIT authorities to keep two parallel accounts, one for the Institute and the other for the Laboratory.

One of the first things that Wiener and Bigelow discovered is that a wave filter that responds very well when the path to follow is smooth will do so erratically when it contains sudden changes in direction; and vice versa, if the predictor goes well for curves zigzagging, its prediction will be very bad for smooth curves. Since one cannot know a priori if a target is going to continue a smooth trajectory or make numerous changes of direction, they concluded that the goodness of a prediction filter could only be evaluated in statistical terms.

Another important aspect related to the problem they were facing is that it involves a continuous feedback process. Indeed, the input of the system — which is the information received by the radars about the position of the target — is subjected to a prediction filter and then the errors committed are measured in comparison to the trajectory of the aircraft received by radar; this information is reentered into the system to reapply the prediction filter, etc.

The statistical factor was addressed by imposing the working hypothesis that the objective's trajectory had a known covariance function. This function will depend, of course, on the aircraft and the pilot, and as they saw it, it had to satisfy the conditions of applicability of the ergodic theorems. In particular, the covariance function should be computable from the values of the path. What's more, Wiener and Bigelow assumed that the calculation could be done simply with the trajectory data in a finite time interval of 10 to 20 seconds, which was the exact time they had to take the shot. Although they were aware that none of these hypotheses fitted with absolute fidelity to reality, they could do nothing else. Furthermore, they trusted that the wave filter they were going to design would be insensitive to the fact that the received signals are finite (and not eternal, as required by the definition of the space S), in the same way that the human ear is insensitive to the transient nature of acoustic waves, thanks to which it can distinguish, for example, the different musical notes.

Wiener and Bigelow worked under pressure. For months, they barely slept, and to cope with it, Wiener took Benzedrine while spending entire nights doing computations. Of course, this excess of activity, together with keeping secret the investigations he was conducting (something that made him very nervous), took its toll on his health and his relationships with the Radiation Laboratory members. He was always irritated and could not present the required reports within the agreed deadline.

Finally, on February 1, 1942, Wiener sent Weaver a 120-pages report with the obscure title "Extrapolation, Interpolation and Smoothing of Stationary Time Series," in which he gave an account of the theoretical results he had obtained. The word "Extrapolation" is the technical name with which mathematicians refer to the problem of predicting the future values of a signal from the knowledge (probably partial) of its past. With "Interpolation," Wiener referred to the problem of estimating the position of the target between two or more known positions. Finally, the expression "Smoothing of Stationary Time Series" refers to the problems of filtering and regularization of the signals. In his report, Wiener outlined a radically new approach to communications engineering. In particular, he merged into a single theory, dedicated to control-engineering problems, two subjects that until then had carried out their respective courses independently: the statistical theory of time series (or stochastic processes) and communications engineering. This would give rise to what was later baptized by Lee (Wiener's Chinese student) as the statistical theory of communication and would mean a revolution in the engineering world.

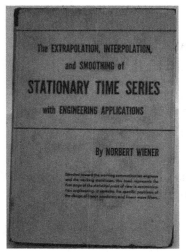

Weaver quickly realized the importance of the document. The first thing he did was classify it as secret. After binding it between yellow covers, he distributed about fifty copies to the engineers and physicists working for the NDRC on related issues, with a clear warning for secrecy. Soon enough, the Wiener report was a cherished document. Those who were aware of it referred to it as the "Yellow Peril" because of the color of the cover and the devilishly complex mathematics it contained.

It must be said that, simultaneously and entirely independently, the Russian mathematician Andrei Kolmogorov (1903–1987) was developing a parallel theory in his country, with the misfortune for Wiener that the articles published by Kolmogorov appeared directly in public. In contrast, Wiener had to wait until 1949 to see his work published, when its secrecy was declassified. On this matter, Wiener would later include the following comments in his autobiography:

> When I first wrote about prediction theory, I was not aware that some of the main mathematical ideas had already been introduced in the literature. It was not long before I found out that just before the Second World War an important little paper on the same subject had been published by the Russian mathematician Kolmogorov in the *Comptes Rendus* of the French Academy of Sciences. In this, Kolmogorov confined himself to discrete prediction, while I worked on prediction in a continuous time; Kolmogorov did not discuss filters, or indeed anything concerning electrical engineering technique; and he had not given any way of realizing his predictors in the metal, or of applying them to anti-aircraft-fire control.

Nevertheless, all my really deep ideas were in Kolmogorov's work before they were in my own, although it took me some time to become aware of this. A series of papers by Kolmogorov and such pupils of his as Krein continued to appear in the Doklady (Reports of the Russian Academy of Sciences), and although these papers still stuck for the most part to the concept of prediction previously developed by Kolmogorov, somewhat narrower than my own, I am by no means convinced that Kolmogorov was not independently aware of the possibility of some of the applications I had made. If that was so, he must have had to keep them out of general publication because of their importance for the military-scientific work of the Soviets. A recent paper by Krein, in which he makes an explicit allusion to my own work in the field of applications, convinces me of this.

I have never met Kolmogorov, and indeed I have never been in Russia, nor have I been in correspondence with him or with any of his school. Thus what I say about him is largely surmise. At an early stage of my work for the United States military authorities, before I had seen Kolmogorov's paper, the question came up whether anybody abroad was likely to be in possession of ideas similar to mine. I said that they would unquestionably receive no particularly ready reception in Germany; that my own friends Cramer, in Sweden, and Lévy, in France, might well have been thinking along similar lines; but that if anyone in the world were working on these ideas it would most likely be Kolmogorov in Russia. This I said because of my knowledge that for twenty or thirty years hardly had either of us ever published a paper on any subject but the other was ready to publish a closely related paper on the same theme.

Unfortunately, the assumptions imposed by Wiener and Bigelow on the prediction problem, although much more flexible than the deterministic methods that had been used to address this problem until then, were excessive. The wave filters were sensitive to the deficiencies caused on the covariance function by the absence of information for time values very far from $t = 0$. This had the effect that, although they had come up with an exciting theory, it was not applicable, with the restrictions imposed by practice, to the particular war problem they were working on. Wiener acknowledged this in his final report and recommended not to continue investing in this project because it would not yield useful results at that time.

However, it must be said that if the digital computer technology developed a few decades later had existed at that time, the work of Wiener and Bigelow would have been applicable to the problem of antiaircraft batteries. In fact, in his 1942 report, Wiener already included the discrete version of the problem, with the corresponding solution, which would give rise to what engineers now call "Wiener filters" (forgetting, perhaps with some intentionality, the work of Hopf), and that it is perfectly adapted for working with digital computers. On the other hand, the incorporation of approximation fuzes made the Wiener predictors highly efficient. Still, these predictors were not favored because their implementation was much more complicated than that of the methods proposed by the other group of researchers, and with the use of fuzes, both methods produced indistinguishable results.

After the war, Wiener would continue to work on problems related to filter theory and, over time, came to consider tackling the much more difficult problem of nonlinear filters, a quest that, according to Lee, took everyone by surprise:

> On December 8, 1949, while staying at the National Cardiology Institute of Mexico, [Wiener] sent a letter to Dr. J. B. Wiesner, who was then the Associate Director of the MIT Electronics Research Laboratory, saying: "I am sending you some information about what I think about non-linear circuits and their treatment. Instruments can be made and Lee can design them. Theoretically, there is nothing new in this, but I think that as an engineering technique it is a hot topic."
>
> Accompanying the letter was a 20-page memoir entitled "The Characteristic Properties of Linear and Nonlinear Systems." About half of the memoir dealt with the subject in general terms, and the Ergodic Theorem. Unfortunately, the "hot spot" was in the second half of the memoir and was not comprehensible to me, so no serious consideration was given to the possibility of me designing the corresponding circuits at the time. The document was patiently studied by several of our colleagues and doctoral students, but none of us came to a satisfactory interpretation. However, we knew that this was an important piece of work and that it should receive continued attention. Fortunately, our combined research effort in this area, which Wiener had initiated during the period 1950–1964, produced more than ten doctoral theses, with numerous practical applications. Wiener's fundamental contributions on this topic are collected in his book *Nonlinear Problems in Random Theory*, in the form of transcribed lessons.

The story of how the people from the Research Laboratory in Electronics (RLE) came to grips with Wiener's manuscript and finally managed to take advantage of it is interesting. In the fall of 1953, A. G. Bose (1929–2013), an MIT electrical engineer of Bengali origin who had arranged to do his doctoral thesis with E. Guillemin (1898–1970) on a topic in conformal maps and their use in electrical networks, returned from a rest trip to the Netherlands ready to start his research. Upon reaching Guillemin's office, he found a note asking him to appear before the director of the RLE. There, he met J. Wiesner (1915–1994) and H. Zimmermann (1916–2007) who convinced him to change the subject of his thesis and join the RLE to study Wiener's manuscript and write a thesis on nonlinear filters under Lee's advisement. Bose did not want to break his engagement with Guillemin, but eventually relented and spent the next ten months studying some notes containing Wiener's 1949 manuscript and some supplementary notes taken by the various people who had read the work. Bose could not fully understand the pamphlet and periodically went to Lee, who could not do much for him because he himself did not understand it. Bose's level of understanding was so low that he did not even bother to go to Wiener since he did not know exactly what question to ask. When he visited Lee, the answer was "Don't quit. Stay in the problem and everything will come." When six months passed, Lee enrolled Bose in an international math conference at MIT and told him that he would have to talk about nonlinear filters there. Bose was terribly scared because he still didn't have a clear idea of the subject, so he spent the next few months studying under enormous stress. But just a couple of weeks before the event, his mind lit up, and he suddenly understood Wiener's insights. His address was a success. Even Wiener was impressed and asked questions. A few days later, Lee went to Bose's office and told him that the attendees had voted and it had been decided that the best talk in the conference had been his presentation of Wiener's theory.

After Bose's conference, Wiener, who used to visit Lee's group sporadically, began to visit them more frequently, until eventually he saw them daily. He always talked to Bose, who, after completing his doctoral thesis and spending a year in India as a Fulbright scholar, joined the Statistical Theory of Communication group permanently as Senior Lecturer in 1956.

One day, after spending a long time visiting Bose, writing equations on the board, and erasing them as soon as he wrote them to continue with others — his usual method in teaching — Wiener wrote one last equation and said, "That's all, Bose: write it down." Bose had only understood a small part of what was explained, and had to confess it to Wiener (who almost could not give credit to what he was hearing). It was then that Bose and Lee had the idea of inviting Wiener to teach a course on nonlinear filters, which they would transcribe in the form of a monograph. While Wiener spoke, they recorded his voice, and each time he finished writing on the blackboard they took a picture of the blackboard and numbered it. In this way, they were able to draw up the text "Nonlinear Problems in Random Theory," which was published in 1958. In 1964, Bose founded a company, the Bose Corporation — specializing in audio equipment — which made him considerably wealthy. With a fortune of 1.8 billion dollars, he was in the middle of the Forbes list of the 400 most important fortunes of the world in 2007. Still, Bose did not leave MIT until 2005.

Hiroshima

There is no doubt that the explosion of the atomic bomb first in Hiroshima and a few days later in Nagasaki greatly shocked the international scientific community. Although it may be ironic, the shock was strong in the United States. Some of those who had been actively involved in the "Manhattan Project" became aware of the appalling outcome of their investigations only after than explosions. There were two very marked types of reactions. There were those, like von Neumann, who supported the use of the bomb and in fact called for continued research on other possible nuclear weapons. The main reason for them was national security. On the other hand, there were those who protested against this new type of weapon of mass destruction. Wiener was evidently on this latter side. For him, the use of atomic bombs was a real shock. Suddenly, the lives of millions of people were in the hands of a few seemingly unscrupulous men. The step they had taken was irreversible and had the effect of making this world progressively uninhabitable.

In all certitude, Wiener was glad that he had not participated in the activities of the Manhattan Project, despite the fact that at one point he had been "called up":

> One day during the Second World War I was called down to Washington to see Vannevar Bush. He told me that Harold Urey, of Columbia, wanted to see me in connection with a diffusion problem that had to do with the separation of uranium isotopes. We were already aware that uranium isotopes might play an important part in the transmutation of elements and even in the possible construction of an atomic bomb, for the earlier stages of this work had come before the war and had not been made in the United States.
>
> I went to New York and had a talk with Urey, but I could not find that I had any particular qualification for solving the special problem on which he requested help. I was also very busy with my own work on predictors. I felt that there I had found my niche for the duration of the war. It was a place where my own ideas were particularly useful, and where I did not feel that anyone else could do quite so good a job without my help. I therefore showed no particular enthusiasm for Urey's problem, although I did not say in so many words that I would not work on it. Perhaps I was not cleared for the problem, or perhaps my lack of enthusiasm itself was considered as a sufficient reason for

not using me, but that was the last I heard of the matter. This work was a part of the Manhattan Project and the development of the atomic bomb.

(...) Be that as it may, while I did not have any detailed knowledge of what was being done on Manhattan Project, the time came when neither I nor any other active scientist in America could fail to be aware that such a project was under way. Even then we did not have any clear idea of how it was to be used. We were afraid that the main use to be made of radioactive isotopes was as poisons. We feared that here we might well find ourselves in the position of having developed a weapon which international morality and policy would not permit us to use, even as they had held the Germans back from the use of poison gas against cities. Even were the work to result in an explosive, we were not at all clear as to the possibilities of the bomb nor as to the moral problems which its use would involve. I was very certain of at least one thing: that I was most happy to have had no share in the responsibility for its development and its later use.

After the bomb blast, Wiener was devastated. He was quick to react. He decided that in no way would he ever share information about his work with governments, especially with the military. Consequently, he would refuse to participate in projects or events that had any military funding. At the end of 1946, he also had the opportunity to make his decision public. Indeed, around that time, he received a letter from an investigator from the aeronautical company Boeing in which he requested from Wiener a copy of his report to the NDRC since he was working on the construction of guided missiles for the American naval forces. Wiener not only refused to provide him with a copy of his Yellow Peril but also refused to send him any other information about his research work and took the opportunity to send an expanded version of his reply to the well-known magazine *The Atlantic Monthly*, which published the letter in its December 1946 issue. That same letter was also published one year later by the Association of Atomic Scientists in its official bulletin. Obviously, Wiener intended to influence the consciences of other American scientists since he believed that

Through non-cooperation, scientists, if united, could paralyze the actions of an irresponsible government.

A Scientist Reveals

The following is the letter written by Wiener in December 2, 1946 to Mr. George E. Forsythe, from the Physical Research Unit Boeing Aircraft Company, and published later in *The Atlantic Monthly*, in January 1947:

Sir:

I have received from you a note in which you state that you are engaged in a project concerning controlled missiles, and in which you request a copy of a paper which I wrote for the NDRC during the war on the subject of prediction and filter theory.

As the paper is the property of a government organization, you are of course at complete liberty to turn to that government organization for such information as I could give you. If it is out of print as you say, and they desire to make it available for you, there are doubtless proper avenues of approach to them.

When, however, you turn to me for information concerning controlled missiles, there are several considerations which determine my reply. In the past, the comity of scholars has made it a custom to furnish scientific information to any person seriously seeking it. Even so, this custom has never been held to apply to information for the use of industrial corporations, and I do not see why we are under any obligation to take a foolishly magnanimous view of the subject, so far as industrial corporations go. But this is a secondary point. The main issue is this: The policy of the government itself during and after the war, say in the bombing of Hiroshima and Nagasaki, has made clear that to provide scientific information is not a necessarily innocent act, and may entail the gravest consequences. One therefore cannot escape reconsidering the established custom of the scientist to give information to every person who may inquire of him. The experience of the scientists who have worked on the atomic bomb has indicated that in any investigation of this kind the scientist ends with the responsibility for having put unlimited powers in the hands of the people whom he is least inclined to trust with their use. It is perfectly clear also that to disseminate information about a weapon in the present state of our civilization is to make it practically certain that the weapon will be used. In that respect the controlled missile represents the still imperfect supplement to the atom bomb and to bacterial warfare. The practical use of guided missiles can only be to kill foreign civilians indiscriminately, and it furnishes no protection whatsoever to civilians in this country. I cannot conceive a situation in which such weapons can produce an effect other than extending the "kamikaze" way of fighting to whole nations. Their possession can do nothing but endanger us by encouraging the tragic insolence of the military mind. If therefore I do not desire to participate in the bombing or poisoning of defenseless peoples — and I most certainly do not — I must take a serious responsibility as to those to whom I disclose my scientific ideas. Since it is obvious that with sufficient effort you can obtain my material, even though it is out of print, I can only protest "pro forma" in refusing to give you any information concerning my past work. However, I rejoice at the fact that my material is not readily available insofar as it gives me the opportunity to raise this serious moral issue. I do not expect to publish any future work of mine which may do damage in the hands of irresponsible militarists.

I am taking the liberty of calling this letter to the attention of other people in scientific work. I believe that it is only proper that they should know of it in order to make their own independent decisions, if similar situations should confront them.

Very truly yours,
Norbert Wiener

The letter undoubtedly caused a stir. But there was no substantial movement of intellectuals to support him. In particular, very few scientists showed their adherence to his thesis. Of course, some of weight did, like Albert Einstein. In an interview, the German physicist would comment that

I deeply admire and approve of Professor Wiener's attitude.
I believe that a similar attitude on the part of all the prominent
scientists in this country would go a long way toward solving
the urgent problem of national security.

But the step taken did not produce a significant change on a large
scale, and it would pose a profound stumbling block to Wiener's later
academic life. The first mess he got into arose when, shortly after his
letter was published, he realized that long ago he had agreed to participate in an important conference that would take place at Harvard
on "High-speed calculating machines." This conference, organized by
H. Aiken (1900–1973) — the engineer who built the succession of
Mark I, II, III, and IV computers — was funded by both Harvard
University and the US Department of the Navy, and Wiener had not
only published his letter but had soon refused to participate in a
research project run in California by a former colleague from MIT
because the results of that research were supposed to have military
use. So, he called Aiken to explain that he would renounce participation in his conference because he considered it inadmissible to reject
certain types of relationships with "the military" and accept others.
The meeting was only two weeks away, and Aiken had no time to
erase Wiener's name from the documentation handed out to con-
gressmen and the press. He just crossed it out. This had the effect
that journalists went to ask Wiener if his absence was related to his
letter in the *Atlantic Monthly*, to which he replied yes. Aiken took
things personally. According to Wiener, he "thought I was involved
in some kind of complot whose ultimate goal was to discredit him and
turn the congress into a public scandal." Given the proud character
of Wiener, they could not settle their differences.

After deciding not to cooperate with the government on research
issues, the problem that Wiener had to face was not simply restricted
to the conflict with Aiken or other similar ones that might arise. The
reality was that in the United States, the analysis of the role played
by scientists in guaranteeing victory in World War II had led the
government to make the firm decision that a good part of military
research resources should be used to finance all kinds of scientific
projects, so much so that most of the money that universities had
to finance basic research projects (mathematics, physics, chemistry,

etc.) and applied projects (engineering) came from military funds. How, then, were they going to interrupt their relations with the army if they were essentially dependent on their money? Thus, Wiener would not only lose research opportunities for economic reasons but, to make matters worse, he ended up immersed in an almost irreversible process of progressive isolation because his colleagues did not want to be related to him for fear of losing part of the generous funding that they depended on for their studies.

In August 1949, the Soviet government exploded a nuclear bomb in Siberia. This, coupled with the seizure of power by the Communists in the Czech Republic and China, the construction of the Berlin Wall, and the start of the Korean War, could be seen as the starting gun for a mad arms race between the governments of the USSR and the USA. This period that served the strange logic of guaranteeing peace through the construction of increasingly powerful weapons would soon receive the curious name of "Cold War." In particular, in the United States, the construction of the hydrogen bomb or superbomb was considered a priority. It was a problem on which between 1945 and 1949 only a few close-to-power scientists had worked (including von Neumann and Teller) but in which actions intensified from 1950 on. In 1952, the United States evacuated a small island in the Pacific and wiped it off the map after dropping over it the first superbomb they had built. The launch was considered a success and a fundamental advance in the Cold War.

On the other hand, in 1950, Senator J. McCarthy (1908–1957) provoked in the United States, with his accusations of "communists" and "Soviet spies," a real paranoia, an endless witch hunt which would affect all the country's institutions including the universities. In particular, President Eisenhower suggested in January 1953 that the government should not allow any communist to teach in the United States. In response to this, the FBI was ordered to conduct an investigation. In April of the same year, Hoover informed Congress that there was evidence of communist infiltration in American universities. As far as our story is concerned, we can mention, for example, that W. T. Martin and Levinson, director and deputy director of the department of mathematics at MIT, respectively, were called to testify before the commission of inquiry on anti-American activities. Both confessed that, years ago, they had belonged to the communist party and had left it after Stalin's coming to power in the

USSR, but they refused to give other names because they knew what this would mean for those affected. Both suffered the consequences of their silence, but they maintained their moral integrity. Another partner of Wiener who was accused of spying activities was Dirk van Struik. When Wiener found out about his friend's predicament, he was in Mexico City, working with Rosenblueth, so there was little he could do to help him. The only thing he could think of (and did) was to send an urgent letter to the new president of MIT, Professor J. R. Killian Jr. (1904–1988), in which he strongly supported the honesty of his friend:

> I know that Struik is a person of the highest character and honesty (...) He has neither the personality nor the intentions of a conspirator (...) If his relationship with MIT suffers, unless there is a far more damming testimony against him (...) I shall regretfully be forced to submit to you my resignation from MIT.

Struik was suspended from employment, but with full pay, for the duration of the investigation that was being carried out on him. He later recovered his former status at MIT Wiener said absolutely nothing to him about the letter, but in time Struik found out what had happened and as he later stated, he believed that he owed Wiener his employment.

Thus, Wiener was affected by McCarthyism in at least two ways. First, his friendship with some of the accused made him suffer for them and defend them to the best of his ability. Second, he was also investigated, and as was the case with many other citizens, he lived in an environment of tension that kept him restless and worried.

In spite of his *Atlantic Monthly* letter and other subsequent actions that could be described as "antimilitarist," if we honor the truth, we cannot claim that Wiener had always been so critical of his government. Indeed, as P. Masani has well observed,

> Wiener's desperate attempts to enlist in WWI, his happy days at the Aberdeen Proving Ground, his swift and affirmative response to Bush's 1940 war work memorandum, and his involvement in military projects during the two world wars, not to mention his concept of duty, rather suggest a pro-military position.

It is to be supposed, therefore, that it was the new situation generated by the use of the atomic bomb that aroused an alarm in him, an inner light in his consciousness that made him reconsider his position in the world, and very particularly, the possible negative implications of his mathematical work. He had always aspired to make some significant contribution not only in the closed universe of mathematics but also in the broader and more ambitious world of technology and science applications. For example, now that he knew the subtle mechanisms that govern control theory, he realized that by applications alone his filter theory had potential that went far beyond the merely technological. It was even very appropriate to think that interesting results could be obtained for areas as diverse as radar theory, medicine, or biology. But during the last years, he had created a good part of his theories under the pressure of work for war and thinking about technical problems of military interest, such as the design of antiaircraft batteries. Although finally his results had not been used in that direction, there was no doubt that sooner or later they would be used for the benefit of the military. What's more, for several years, his work had been only available to a very small group, shielded by military secrecy. Was that what he wanted to do with his life? Becoming one more piece, probably a piece of great utility, in the military machinery of the State? Should he be submissive and put all his knowledge and skills as a researcher at the disposal of his government, which was now immersed in a stupid Cold War with the USSR whose end could not predictably be peace? Definitely not. If he was a true intellectual, he could not submit to the whims of the politicians or the military, whom he increasingly distrusted and whom he considered in reality to be unprepared for the decisions they had to make. An independent intellectual analyzes his surroundings with critical rigor, and if he deems it appropriate, exposes his analysis so that citizens are informed and warned of the risks and dangers that lie in wait for them.

Starting in 1948, Wiener decided to write several books devoted to the general public, in which he would explain his particular vision of science and the important relationships that exist between science and society.

An Elephant in the Room: The Militarization of Science in the USA

As much as Wiener protested publicly about the increasingly important role that the military had in the control of public funding of
American university research, his efforts were doomed to failure. The
military — as it would be seen later — had come to stay. After
the Second World War, many other reasons such as the Cold War,
the Korean War, the Vietnam War, the globalization of communications and information, and the fight against terrorism would come to
maintain the military influence on financing the universities.

For long years, the military would influence the choice of subjects
to which attention should be paid in science and engineering. Many
universities, led by MIT, would soon acquire more and more money
from military funds, developing a significant dependence on them.
As early as 1962, the physicist A. Weinberg complained that it was
complicated to distinguish "whether the Massachusetts Institute of
Technology is a university with many government research laboratories appended to it or a cluster of government research laboratories
with a very good educational institution attached to it."

In *The cold war and American Science*, S. W. Leslie contributed
much data that evidenced the dependence of MIT on military funds
not only during World War II but also decades later. For example,
in 1968, MIT received funds of military origin just below what the
Department of Defense (DoD) was paying the company TRW on a
giant missile project and above what was paid to Thiokol Chemical,
a major chemical company that produced rocket and missile fuel.
What's more,

> Most of [the money] went into MIT's interdepartmental lab-
> oratories, the Research Laboratory of Electronics (RLE), the
> Laboratory for Nuclear Science and Engineering, the Lincoln
> Laboratory, and their many spinoffs. These became the center-
> pieces of postwar MIT, the places where the important research
> was done, graduate students trained, widely adopted textbooks
> written, and future Institute leaders groomed. These laborato-
> ries also provided the benchmarks, the models, and the faculty
> for MIT's many competitors and imitators.

The RLE was the first of the multidisciplinary laboratories created by MIT in which significant advances were made in matters of military interest. And a critical aspect in the research that would be developed there was precisely the theory of communication. It was also the mold in which many other laboratories supported by military funding were created. Where once the support of industry and philanthropic foundations was sought, there was now a far more reliable source of financing: the seemingly inexhaustible federal defense resources.

Of course, this was a real boost for research, especially in physics and engineering, giving rise to what was soon called Big Science, a term that refers to the use of large structures or facilities, such as particle accelerators, with the institutional support from one or more governments. Some physicists were delighted and came to boast that there was unlimited money to advance their projects — as long as you devoted yourself to particle physics — and the same happened without doubt in other environments that were also rewarded for their contributions to military projects. However, voices also emerged, such as that of the astronomer Harlow Shapley (1885–1972), warning of the situation:

> Those who were worried about domination of freedom in American science by the great industries, can now worry about domination by the military.

In addition, the dependence on military funds — which, over time, would reach unimaginable heights in the case of MIT — also had a high risk because the money could suddenly disappear if there were a sudden turnaround in Washington's policy, and that was something that seriously worried the academic authorities.

This whole process of "militarization of science" had begun in 1945, when the Office of Naval Research (ONR) was created, with the intention of forming a group of scientists who would be prepared to support military projects in case of a new conflict. In 1946, this office became by law permanent, and from that moment on, new organizations began to emerge, such as the Atomic Energy Commission (AEC), the National Science Foundation (NSF), the Army Research Office (ARO), or the National Aeronautics and Space Administration (NASA), among others. Of all these organizations, only the NSF was purely civilian in character, and its creation was motivated precisely by the desire to avoid overreliance on military science research. It was only three months after its creation that the Korean War broke out, which had the effect of intensifying the activity of the military agencies, and the NSF fell into the background. In "El poder de la ciencia," the Spanish physicist and historian of science J. M. Sánchez-Ron argued that:

> What the US military wanted was to control a substantial part of the nation's scientific potential; to maintain in a balanced, and not a too much conspicuous way, a network of scientific and technical facilities and personnel that would serve military purposes. These purposes did not necessarily imply involving these professionals in clearly warlike investigations, although this was also done, of course. The US military did not fall into the error of believing that they should only fund research aimed at producing new weapons. They recognized that it was necessary to maintain permanent relations with the academic world, since if not, they ran the risk of losing everything that had been painstakingly achieved during the war. This required a complex and refined scientific policy and do not alienate the professionals of science.

Furthermore:

> There was neither naïveté nor improvisation in the support that the armed forces gave to science as of 1945.

In fact, with the data currently available, it is indisputable that the Department of Defense got what it wanted, and it got it very quickly. For example, a report from that Department dated 1951 showed that 70% of the research time of physicists working at 750 American universities was devoted to military research. The result of this type of research is, to a large extent, numerous inventions that currently have very diverse uses, such as lasers, masers, radar, nuclear energy, and even — to give an example of daily use — Teflon, which originated as a microwave dielectric material. This has been the main argument put forward in favor of the militarization of science by different academic leaders over the years. However, the truth is that many of these technologies could have also been developed from premises outside of war. Although military research has made important advances in creating civilian and socially useful technologies, this does not mean that without military funding the same technology would have been impossible or that, for example, it would necessarily have arrived later. Similarly, the total elimination of military research could not be seen as a guarantee of faster progress in civil applications either.

In the case of MIT, there were several decisive moments in which specific commissions were created to analyze the entire institution. This included the participation of industry and the military in funding not only research laboratories but also in the creation and award of grants, research projects, etc.

The first of these commissions, promoted in 1947 by the chemical engineer W. K. Lewis, alerted in his conclusions that "the energy and interests of some of our most talented colleagues" had been diverted away from creative teaching and research towards narrowly focused, "income-producing" projects and that MIT's finances were too dependent on federal funds. They concluded by recommending "a serious effort to stabilize our over-all size within limits that will avoid purposeless dissipation of our abilities and resources." In addition, the commission required creating a faculty of humanities that would allow students to become aware of the social relevance of the work being done at MIT, advice that was followed immediately.

However, as D. Kaiser reminds us in *Becoming MIT: Moments of Decision*, the commission's claims related to military funding were not addressed:

> Amid the flush and exuberance of runaway growth, few heeded the Lewis committee's warnings. Only after national priorities shifted and growth began to falter two decades later did many on campus pause to take stock of the costs of exponential expansion. Often dubbed the "Pentagon on the Charles" by its critics — owing to the Institute's close relationships with military patrons — MIT had become an ungainly elephant by the 1960s, stretched and swelled like few other universities in Cold War America.

In 1970, after detecting a significant drop in federal revenue from 1965 on, a second commission was created, this time led by the mathematician K. Hoffman. The situation was no longer one of euphoria, but rather a turbulent time, with student protests over the Vietnam War that were dissolved by the anti-riot units of the police:

> Starting in the mid-1960s, Department of Defense officials began to reconsider whether open-ended basic research had produced the best return on their investment. The Mansfield Amendment to the Defense Appropriations Act of 1970 legislated what military planners had already been leaning toward on their own: it restricted military spending on the nation's campuses to projects of direct military relevance. There would be no more blanket grants to universities for the chief purpose of drumming up scientific recruits. Meanwhile, the fiscal demands of the escalating war in Vietnam, combined with the first glimmers of "stagflation" — rising inflation and stagnant economic growth — led to massive cuts in federal spending on education. MIT, having become so dependent on federal grants, felt the shockwaves first. Budgets quickly began to slide. MIT faced a budget shortfall of $10 million between 1971 and 1973, its first deficit in decades. Enrollments likewise reached a plateau and then began to slip.

Hoffman's report, entitled *Creative Renewal in a Time of Crisis*, insisted on taking into account some of the main aspects of the previous commission's report. It now openly criticized the military's influence in the development of the too-focused curriculum and the emphasis on resolution of highly technical problems, leaving aside the search for creativity and judgment.

Years later, this same complaint was ratified as a generalized practice in American universities, when in 1987 C. Barus published in *IEEE Technology and Society Magazine* an article entitled "Military influence on the Electrical Engineering Curriculum since World War II," in which he presented very solid evidence that "military sponsorship has not only hastened postwar progress, as generally agreed, but has shaped it." Indeed, as he argued in his article, it is enough to apply a little common sense to realize that:

> Professors teach what they know. They write textbooks about what they teach. What they know that's new comes mainly from their own research. It is hardly surprising, then, that military research in the university leads to military-centered undergraduate curricula.

The reality described by Barus in 1987 has not changed substantially. To a large extent, MIT — like many other universities and research institutes — is still under the military's influence and that of the biggest technology companies. Wiener's warning has had little effect. For many professors and researchers, there is a feeling that they are working in a room where a huge elephant controls everything. Still, no one in the room mentions the elephant, and everyone does their job as if they did not see it, as in Andersen's tale *The Emperor's New Clothes*, where no adult dared to claim that the emperor was nude.

Chapter 6

Cybernetics: The Return to Philosophy

From Mathematician to Scientific Popularizer

The publication of his letter in the *Atlantic Monthly* placed Wiener at the center of the American scene for the second time in his life. The media showed interest in this somewhat unusual looking mathematician, who had been a child prodigy at the beginning of the century, and gave him a voice. But that would not have lasted long if Wiener had returned to regular mathematical work, again concentrating on some purely scientific problem and sending the result of his efforts to some obscure research journal. Instead, he decided to devote a good part of his time to explain the state of science and technology of the moment, and, projecting himself into the future, he tried to predict its benefits and its dangers. Wiener wanted to alert the average citizen of the new state of affairs, which was so disturbing for him.

To achieve his goals, he decided to become a popularizer and embarked on a path that led him back to philosophy, and he made this step precisely when he was in the final stretch of his academic career.

To those who knew his mathematical work, he was famous for his somewhat chaotic way of writing articles, sometimes leaving a profound result unproven or boring the reader with seemingly insignificant details. Nonetheless, Wiener achieved considerable success as a disseminator. He retained the media's attention, which increased over time as the different books he wrote for the general public appeared.

Cybernetics and *The Human Use of Human Beings* covers.

Among these, the two that are his best-known works that have had the greatest impact on the science of the 20th century are the two versions of *Cybernetics*. The first saw light in 1948 and is entitled *Cybernetics or Control and Communication in the Animal and the Machine* (we refer to it from now on as simply *Cybernetics*). The second, published in 1950, is entitled *The Human Use of Human Beings: Cybernetics and Society*, and we cite it as *The Human Use of Human Beings*.

Although Wiener claimed that they were two versions of the same book, the truth is that they are very different works. Both were intended for the general public, but reading the first one requires facing complex (and often incomplete) mathematical developments, which the author uses to explain the basic principles on which his new discipline was founded. At the same time, the second book has a deep philosophical and social focus and does not contain formulas. Interestingly, it was the first of these works that gave him worldwide fame, and many more translations of it have been published than of the second. However, in *The Human Use of Human Beings*, we find a more accessible Wiener who transmits his ideas better and more accurately.

Cybernetics: Origins of the Term and Later Use

It is no coincidence that the origin of the word "cybernetics" is in the Greek word Κυβερντης, which means "helmsman." Plato already used the term in his *Republic* with the meaning "rule" (in the sense of "govern"). Wiener's translation of the classical "kibernetes" to the modern "cybernetics" demonstrates his passion for languages and philosophy.

There is no doubt that the modern father of this discipline and the person who started to use the term with the definition accepted today in our dictionary was Wiener in 1948, with the publication of *Cybernetics*. However, and although this book is usually cited as a seminal work in the subject, most of the primary sources of cybernetics appear intertwined with their own fields of application, which include, among others, general systems theory, bionics, the theory of self-organizing systems, evolutionary systems, and, finally, artificial intelligence.

Wiener could have chosen another word to designate this unifying theory, and in fact he initially thought of the word "angelos," which is the Latin version of "messenger." The closeness to the "angel" (messenger of God) made him quickly abandon the idea to avoid any confusion. When asked why he chose this word, Wiener just replied, "I couldn't think of another way to call this."

If we want a modern and updated definition, the dictionary of the Royal Academy of the Spanish Language gives us the correct meaning, in the precise sense in which Wiener defined it:

> Study of the analogies between the control and communication systems of living beings and those of machines; and in particular, the [study] of the applications of biological regulation mechanisms to technology.

And, the *Cambridge Business English Dictionary* tells that Cybernetics is:

> The scientific study of how information is communicated in machines and pieces of electronic equipment in comparison with how information is communicated in the brain and nervous system.

As can be seen, both definitions are practically a carbon copy of the complete title of Wiener's book, updated with the meaning that it acquired shortly after with the progressive development of the discipline thanks to the contributions of Wiener and others.

Having framed what Wiener understood by cybernetics, it is worth highlighting that currently this term can easily be misleading. In popular culture, cybernetics is associated with robotics and the concept of cyborg (a neologism for a hybrid between human and machine, popularized by Clynes and Kline in 1960). The current degeneration of the term, often through the use of the prefix "cyber" to denote any activity related to electronics or digital communications, is far from its initial meaning. Thus, nowadays, we have the terms "cyberspace," "cyber-coffee," "cyberpunk," or even "cyberterrorism." All these words are derived from "cyberspace," invented by William Gibson in his 1984 novel *Neuromancer*, which in turn is derived from cybernetics. It seems that this new prefix term has ended up displacing the meaning of the original one.

The Road to Fame

In 1947, Wiener traveled to Nancy (France) to attend a conference on harmonic analysis organized by his friend S. Mandelbrojt. The meeting was attended by numerous leading mathematicians, including H. Bohr (1887–1951), T. Carleman (1892–1949), A. Ostrowski (1893–1986), and M. Plancherel (1885–1967) (whom, as we know from his work on Fourier analysis, Wiener deeply admired). But Wiener especially remembers L. Schwartz (1915–2002), who was closing his theory of distributions precisely at that time. In this particular instance, Wiener would comment,

> Schwartz was active along lines very similar to my own. He had generalized still further the field which I had already treated in my Acta paper on generalized harmonic analysis. He reduced it to that highly abstract basis which is characteristic of all the work of the Bourbaki school to which he belonged.

As was customary for Wiener, he took advantage of this trip to Europe to visit his friends. In particular, he traveled to England where he met Haldane, who was newly married to his second wife, and he also stopped by Paris.

There, he met E. Freymann, a Mexican who had a small bookstore in front of the Sorbonne. That bookstore was a meeting place for many intellectuals, and it was easy to engage in all kinds of conversations. Wiener found in Freymann "the most interesting man I have ever met." In his biography, he described him as follows:

> Freymann, who, alas, has just died, was a Mexican, and he first came to Paris as a cultural attaché in the Mexican diplomatic service. One of his grandfathers was a retired German sea captain, who had made his home in the region of Tepic, on the west coast of Mexico. The other grandfather was a Huichol Indian chieftain from the same region. My friend's two grandmothers were both of Spanish origin. Freymann, who kept a drab little bookshop opposite the Sorbonne, where every now and then one of the scientific or other intellectual notables would drop in, delighted in relating to me the way each of his two grandfathers tried to capture him from the influence of the other, the one telling him always to be European, and the other reminding him that he was an Indian.

Apparently, the conversation between Wiener and Freymann was very fluid. Although they initially spoke only about Mexico and typical family issues, the talk changed from topic to topic until they reached Wiener's scientific work. Wiener certainly explained his new concerns and in particular his concern about the "secrecy" with which many scientific investigations — especially those related to military issues — were being carried out. Furthermore, at that time, Wiener had turned his interests into the many applications that the theory of prediction and the theory of wave filters had in matters as diverse as physiology, medicine, or economics.

There is no doubt that Wiener's interest in Mexico (due in large part to his friendship with Rosenblueth, with whom at that time he was actively collaborating and whom he had visited at the Mexican Institute of Cardiology since 1945) was the meeting point between him and Freymann. Indeed, Wiener was impressed by Mexico from his first visit, which he remembered as follows:

> Almost from the moment I crossed the frontier, I was charmed by the pink-and-blue adobe houses, the bright keen air of the desert, by new plants and flowers, by the indications of a new way of living with more gusto in it than belongs to us inhibited North Americans. The high, cool climate of Mexico City, the vivid colors of the jacaranda blossoms and the bougainvillaeas, the Mediterranean architecture, all prepared me for something new and exciting. The many times I have returned to Mexico have not belied these first impressions. It will be a sad day for me when I come to feel that I have no further chance to renew my contacts with that country and to participate in its life.

As discussed in Chapter 4 of this book, Wiener met Rosenblueth in 1933, and they were close friends since then. The way they became intimate was through participation in a seminar on the philosophy of science organized by Rosenblueth at Harvard, which was known as "The Philosophy of Science Club."

> In those days, Dr. Arturo Rosenblueth [. . .] led a monthly meeting dedicated to the discussion of the scientific method. Most of the participants were young scientists from Harvard Medical School, and we met for dinner at a round table in Vanderbilt Hall. The conversation was lively and had no restrictions. It was not a place to seek approval or to hold a position. After dinner, someone — either outside of our group or a guest — would read a scientific topic, which, in general, was about methodology, or at least had to do with it. The exhibitor had to face sharp criticism, well-meaning but ruthless. It was a perfect catharsis for undefined ideas, insufficient self-criticism, exaggerated confidence, and pomposity. Those who did not accept that environment did not return and among the regulars we found several who felt that these reunions were an important part of our scientific development.

Furthermore, in 1943, Wiener had resumed his relationship with Rosenblueth, this time working together on a mathematical model, based on the theory of control, of "intentionality." Later, starting in 1945, they developed a series of works on the physiology of the heart and nervous system.

Arturo Rosenblueth (1900–1970)

Arturo Rosenblueth (1900–1970). Courtesy: Public domain, Wikimedia Commons.

He was a Mexican physiologist and one of Wiener's closest collaborators during his work in cybernetics. Rosenblueth was born in Chihuahua. After a difficult childhood and receiving a scholarship to study in Europe, he obtained a medical degree from the University of Paris in 1927.

In 1930, he got a scholarship to work in the department of physiology at Harvard. There he developed, together with W. B. Cannon (1871–1945), several papers in the field of nerve transmission. He specifically studied the problems of chemical communication between nervous elements and between them and effectors, contributing significantly to the development of mechanical and electrical recording techniques.

In 1933, thanks to the mediation of the physicist M. Vallarta (1899–1977), Rosenblueth met Wiener, who would be "amazed" at the seminar that Rosenblueth taught at Harvard about the scientific method. A close friendship accompanied the scientific collaboration. A solid proof of this is the dedication that Wiener writes in Cybernetics: "To Arturo Rosenblueth, a fellow scientist for many years."

In 1944, and due to the strong restrictions applied by the United States government, North American citizenship was an essential requirement to obtain a permanent job position at the University of Illinois. After he refused American nationality, and despite the letters of recommendation that among others McCulloch (who considered him "the best physiologist in America") and Cannon sent to the government, Rosenblueth returned to Mexico, where he was appointed Director of the National Institute of Cardiology.

Once there, he continued his work with Wiener under the auspices of the Rockefeller Foundation. A year before his return to Mexico, he wrote, together with J. Bigelow, the work entitled *Behavior, Purpose and Teleology*, which some place as the starting point of cybernetics.

Near the end of his life, he published *Mind and Brain: A Philosophy of Science*, where he summarized his views on the philosophy of science and some of his scientific achievements in physiology.

What Wiener could not know was that Freymann was the son-in-law of the owner of the publisher Hermann and that the owner had recently passed away. Freymann was the only relative willing to take over the publishing company. So, Wiener must have been surprised when, after telling him about his scientific concerns, Freymann proposed that he write a monograph dedicated to these questions. Moreover, Freymann asked that, if possible, the book should be aimed at a broad audience. The book would appear in the series of scientific publications of Hermann. Wiener accepted immediately, and "they closed the deal by drinking chocolate in a local bakery."

During the rest of his stay in Paris, Wiener did not bring up the subject again. He also did not work on the book during the few days of his return to the United States, as he had to prepare for a new visit to Rosenblueth at the National Institute of Cardiology. However, when he was established in Mexico, Wiener got down to business. He spent his free time (especially the first hours of the day after waking up, when he felt fresher) writing the manuscript. It turned out to be a compendium of a good part of his previous research in prediction theory, feedback and physiology, statistical mechanics, etc., oriented to the general public, with a broad philosophical and unifying background, one of the ultimate objectives of which was to give a modern vision of the mechanisms of the human mind.

Wiener organized the book in the following way. He wrote a long historical introduction where he recounted the genesis of the new scientific discipline that the book was going to deal with: cybernetics. Next, he dedicated the first chapter to speculate on the differences between Newtonian time and Bergsonian time, which he considered vital to define the framework in which cybernetics develops. After this, he wrote a chapter dedicated to groups and statistical mechanics, another about time series, information, and communication, then a chapter about feedback and oscillation, another on computers and the brain, a small chapter on gestalt and universals, another on cybernetics and psychopathology, and finally the book ended with a chapter devoted to information, language, and society.

When the manuscript was finished, Wiener sent a copy to Freymann, who immediately accepted it for publication. But Wiener also showed the book to the MIT authorities, with the idea of getting their opinion. The MIT publisher was genuinely interested in publishing the monograph in the United States, but this posed a moral problem for Wiener, as he had already transferred the rights to Freymann. In the end, the book was published in both countries simultaneously.

To everyone's surprise, including Wiener himself, the book was very warmly received. Although there were indeed some harsh criticisms, in which the excessive technical details of the book were emphasized, as well as the appearance of various mathematical errors and misprints, the truth is that very soon it was considered an important essay, with a reach far beyond the merely technical. As proof of this social impact, we can see what on December 19, 1948 an extensive review in the *New York Times* commented:

> *Cybernetics* [...] has brought this new science to public attention. Until now, cybernetics was known only to a small circle of the elite among which it has been considered something like a new revelation. Some of Dr. Wiener's most enthusiastic devotees believe that his further development will lead to a revolution in the understanding of the human mind, and human behavior, both normal and abnormal, comparable to the revolution in our understanding of the physical world initiated by quantum theory and relativity.

On the content of the book and the impact on other scientific areas, this review added:

> *Cybernetics* offers a new approach to the study of the human mind and behavior, based on a comparative study of the electrical circuits of the nervous system, and those that are part of the highly complex computing machines. In view of this relationship, it promises to become an object of vital interest to psychologists, physiologists, electrical and radio engineers, psychiatrists, mathematicians, anthropologists, sociologists and philosophers.

Another impressive praise was published in 1950, in the American *Saturday Review of Literature*, where the following was stated:

> It is impossible for anyone who is seriously interested in our civilization to ignore [Wiener's *Cybernetics* text]. This is a must-read book for those who work in any branch of science.

In a second edition, published in 1961, Wiener corrected the misprints and in addition added two new chapters, entitled *On Learning and Self-Reproducing Machines* and *Brain Waves and Self-Organizing Systems*.

On the other hand, long before the publication of this second edition, Wiener in 1950 published *The Human Use of Human Beings*, that he considered a new version of *Cybernetics* in which mathematical technicalities disappear, and consequently whose public reach was much broader. Of that book, he would publish two versions. The second one, that appeared in 1954, was more readable but lost some of the freshness and sharpness of the first. Furthermore, in 1949, Wiener's wish that his work for the NDRC be declassified was satisfied, and he published a revised version of his *Yellow Peril*. In the words of G. T. Raisbeck, who was the person in charge of preparing the manuscript for publication and who at that time was working for Bell Laboratories, the book had a strong impact among engineers:

> In 1949 people began to apply Norbert's work to signal coding and detection, and from then on, you had to use those methods, because they were much more powerful than what had been used until then. That's when Wiener was first recognized by the engineering community beyond MIT, at other reputable electrical engineering schools.

Thus, 1949 was the year in which Wiener returned to the American scene, no longer as a child prodigy or a first-rate mathematician, but as a true genius, as the creator of a new scientific paradigm. Moreover, this recognition came simultaneously from the lay public and the experts, mathematicians and engineers who worked in the most prestigious research institutions in the country.

> Following his debut on *Time* and *Newsweek*, Wiener's burly presence and his predictions appeared regularly in one-page photo articles on *Life*. His work became the subject of long articles in *Fortune*, *The New Yorker* (...) The French newspaper *Le Monde* published an article on *Cybernetics* and the response to the book was especially strong in Sweden. In 1949 Wiener was invited to deliver the Josiah Willard Gibbs Lecture at the annual meeting of the American Mathematical Society. In 1950 he participated in the ICM, which was held that year at Harvard.

Wiener would not be content with the publication of these books. Now that he had become a public figure, he thought it might be interesting to write his autobiography, which he did in two phases, publishing *Ex-Prodigy* in 1953 and *I am a Mathematician* in 1956. Furthermore, in 1954, he wrote the essay *Invention. The Care and Feeding of ideas*, but then decided not to publish it for ethical reasons. According to Heims, it is quite possible that he thought his manuscript was excessively optimistic, and he feared that such optimism encouraged the creation of new inventions by those whose goals he most detested. In 1959, he published the novel *The Tempter*. In 1963, just a year before his death, he published a philosophical essay titled *God and Golem, Inc. A Comment on Certain Points where Cybernetics Impinges on Religion.*

McCulloch and Pitts: Other Cyber Pioneers

Cybernetics was not based solely on Wiener's ideas. To understand its contents, as well as the impact it had, it is necessary to review the scientific environment in the 1940s, since it was precisely at that time when various groups of people with apparently different interests began to make developments in the same direction indicated by Wiener in his book. In addition to that of Wiener, Rosenblueth, and, perhaps Bigelow, the most influential group was formed around McCulloch and Pitts. Between all of them, and with the virtual collaboration of other scientists such as John von Neumann, the birth of the "cybernetic movement" was achieved.

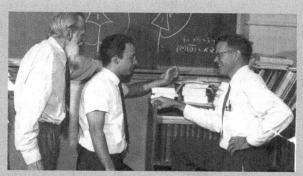

Warren McCulloch with Louis Sutro and Roberto Moreno.
Courtesy: Roberto Moreno.

Warren Sturgis McCulloch (1898–1969). He was born in Orange (New Jersey, USA). He completed an undergraduate degree in philosophy and psychology at Yale, graduating in 1921, and studied a master's degree in psychology at Columbia University, obtaining his degree in 1923. He received his doctorate in 1927 from the College of Physicians and Surgeons of New York, taking an internship at Bellevue Hospital in New York. After practicing as a doctor, he decided to begin his academic career at Yale University, in the neurophysiology laboratory, where he worked in physiology under the supervision of Dusser de Barenne. In 1941, he joined the department of psychiatry at the University of Illinois, Chicago, where he remained until 1952. There, he met Walter Pitts, a young logician with whom he worked on mathematical models of neural activity.

In 1952, he moved to the MIT electronic research laboratory, where he applied the developments conceived in the previous stage of his career to create a neural network model to examine the visual perception system in a frog's eye and the information processing in the brain.

He founded the American Society for Cybernetics and was its first president (from 1967 to 1968).

Multifaceted, McCulloch also did some work in construction engineering and developed a certain poetic talent. Along with Walter Pitts, he is considered the father of neural networks and one of the pioneers in artificial intelligence.

Logician Walter Pitts (1923–1969) in 1954 at a time when he was a research associate and enlisted as graduate student at MIT in Boston. Courtesy: Public domain, Wikimedia Commons.

Walter Pitts (1923–1969). He was born in Detroit (USA). A child prodigy, at age 12, he read Bertrand Russell's *Principia Mathematica* and sent him a letter with a series of corrections to a troublesome issue that appeared in the first volume. Russell invited Pitts to study in England as a token of appreciation, although Pitts declined the offer because he was too young. However, he decided to become a logician.

At just 15 years of age, he ran away from home and showed up at the University of Chicago so he could follow Russell's lessons. There, he met Rudolph Carnap and presented him with an annotated version of his last book on logic, *The Logical Syntax of Language*, with suggestions and corrections made on the book itself. Pitts had not introduced himself, which resulted in Carnap looking for him for months to offer him a (minor) job at the university. Pitts at that time was homeless and unemployed.

Sometime later, McCulloch arrived at the University of Chicago and in 1942, invited Pitts to live with him in his house where in the afternoons the two collaborated. Pitts co-authored with McCulloch a seminal work on the subject of neural networks, in which the question arose as to whether the nervous system is a universal computer in Turing's sense.

Finally, Pitts managed to graduate in mathematics. In 1951, Wiener convinced J. Wiesner (1915–1994), then director of MIT's Electronics Research Laboratory, to hire several neurophysiologists, including Pitts, Lettvin, McCulloch, and P. Wall. Pitts wrote a thesis on the properties of neural networks connected in three-dimensional space. Lettvin would describe him as follows:

> Without a doubt, [Pitts] was the genius of the group ... When you asked him a question, you got a complete textbook for an answer.

On the other hand, he was also described as an eccentric who consistently refused to make his name publicly available. In particular, he refused to obtain a more advanced degree or an official position at MIT simply because he did not want to sign his name.

Shortly after, when Wiener broke his relationship with McCulloch and "his boys," the person who took it the worst was Pitts. He lost all his interest in research, burning the manuscript of his thesis, and gradually became isolated. He died in 1969 aged only 46, from complications of a disease derived from his alcoholism.

As it could not be otherwise, if one of the fundamental interests of cybernetics was to provide a reasonable mathematical model of the brain, there was no choice but to include mathematical logic and the new-born engineering associated with electronic computers —

largely promoted by von Neumann — as an important part of the new science. In particular, this included the recent developments of A. Turing (1912–1954) on universal machines and A. Church (1903–1995) on computability.

Indeed, this was how McCulloch and Pitts entered the scene, offering a theoretical model of the functioning of neurons, which would establish a link between abstract mathematics and neurology.

In 1943, McCulloch and Pitts had already accumulated a year of intense work. They had written an article that would be critical of the development of the theory of neural networks, and that would win the admiration of the scientific community. In that paper, two subjects that were considered independent until then were related, binary logic and brain activity. The work, which they titled "A logical calculation of ideas immanent in nervous activity" and published in a biophysics journal, presented a very elementary neuron model in which the inputs are added through a sum with real weights and the output is binarized using a step function. McCulloch and Pitts tried to explain how the brain can produce highly complex patterns using many interconnected basic units.

The McCulloch–Pitts Neuron

The inspiration for the McCulloch–Pitts neuron model stems from the characteristics observed in brain neurons. In these cells, two types of connections can be distinguished, the dendrites and the axon. Dendrites are a series of branches through which inputs are received from other neurons. The axon is a prolongation of the body with terminations at the synapses. These are the chemical connections between the end of the axon of a neuron and the dendrites on another. This is where communication occurs.

Following this analogy, McCulloch and Pitts consider the neuron as an elementary computing unit, which receives a series of inputs from other neurons, operates on them in some way, and produces an output. Moreover, both the input and the output are represented as one of the numbers 0 or 1, since each neural connection only has two possible states: active, which is characterized by a 1, or inactive, which is represented by a 0. The inputs and outputs are henceforth binary.

For a neuron with n binary inputs, they thought that a good way to combine these inputs is to give each of them a real weight and then binarize the result. This is equivalent to calculating the scalar product of the binary vector of inputs by a vector of real weights, which is a characteristic of each neuron. The output of the neuron as such is a real number, which must be binarized. To do this, McCulloch and Pitts chose to place the Heaviside step function at the output, with a certain threshold. If the sum of the weights of the active inputs exceeds a specific value (which they call the threshold), the output of the neuron is 1. Otherwise, the output is 0.

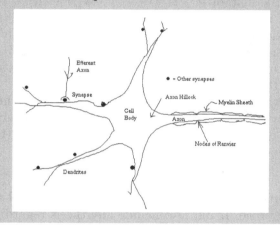

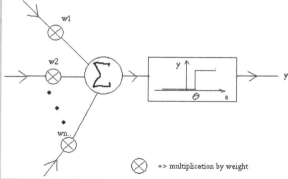

Scheme of a neuronal cell and function of the McCulloch–Pitts neuron.

The formal mathematical expression of these ideas would be as follows: We will say that a formal neuron (or module) is a function m of the binary variables $x_1(n), \ldots, x_k(n)$, which represent the state of activity or inactivity of each of the dendrites of the neuron at the instant n, where $n = 0, 1, 2, \ldots$ is a non-negative integer. Moreover, this function satisfies that

$$m(x_1(n), \ldots, x_k(n)) = 1 \text{ if } w_1 x_1(n) + \cdots + w_1 x_1(n) \geq \theta$$

and $m(x_1(n), \ldots, x_k(n)) = 0$ otherwise. To this, we must add that the value $m(x_1(n), \ldots, x_k(n))$ represents the output of the neuron at the instant $n + 1$.

The real numbers $w_1, \ldots, w_k$ and θ are characteristic of the neuron under consideration, and do not change with time. If $w_k < 0$, then the corresponding input will have an inhibitory function. On the other hand, θ is the value that represents the activation threshold of the neuron.

Finally, the output of the neuron can be used as the output of the system, or as the input of another neuron, giving the possibility of forming a complex network of neurons (that is, a neural network).

The importance of the neural model was theoretical and had no biological relevance. It did not include any thesis regarding the true physiology of the nervous system. Instead of that, starting from a simplified version of the real behavior of the neurons where the biochemistry of the process was completely left aside, a mathematical model for neural networks was provided. Then, it was shown that with this model any logical function of its inputs (such as AND, OR, NOT, or XOR) can be realized. Thus, they showed that the model is equivalent to a Turing machine or, expressed in another way, that any calculable function (in Turing's sense) can be built with a McCulloch–Pitts neural network. Today, the McCulloch–Pitts neuron model is also known as a "logical threshold unit" or LTU.

McCulloch's conclusions were blunt: The first was that the brain performs "logical" thinking, and the second was that any brain process can be described by logic.

From 1943, McCulloch and Pitts, eager to demonstrate the useful-ness of their discovery, embarked on a practical application: making their network of neurons capable of performing a visual recognition task. In the 1947 paper "How we know universals: the perception of auditory and visual forms," they developed a theory about how a set of basic neurons could perform a task of statistical induction, or seen from the point of view of learning, of abstraction. For example, if we had a set of images of lemons, is there a way to "learn" or "abstract" the concept of lemon from these examples?

The inspiration for this second work arose during the first Macy conference, held in 1946, as a result of a discussion proposed by the psychologist Heinrich Kluver, in which the classic problem of perception or "Gestalten" (that is, the problem of how we perceive shapes) was considered.

The Macy Conferences

These were a set of ten meetings of scholars from various disciplines interested in the functioning of the human brain, which were held in New York between 1946 and 1953. The Josiah Macy, Jr. Foundation paid the organization's expenses (Macy was a philanthropist who died in the 1930s from yellow fever, whose daughter created a foundation for medical research and gave the name to these meetings). The Foundation was interested in supporting medical research and the transfer of ideas between different subjects. McCulloch was the person in charge of the organization itself. The ultimate objective of these meetings was to work on the development of "a general science that would describe how the human brain works."

In 1942, long before the first Macy conference took place, there was a small meeting in which McCulloch met Wiener and became acquainted with his work with Rosenblueth and Bigelow. Thus, although the first official conference was prior to the publication of *Cybernetics*, Wiener's ideas together with those of Bigelow and Rosenblueth set the initial direction of the meetings. Not surprisingly, the series' inaugural lecture was entitled "Feedback mechanisms and causal circular systems in biological and sociological systems." Wiener, Rosenblueth, and Bigelow were part of the "core group," which organized for the most influential scientists to attend the meetings. In addition to the three just mentioned, this core group also contained McCulloch, Pitts, W. R. Ashby (a psychologist, who would later be considered a key figure in cybernetics), H. von Foerster (electrical engineer), and John von Neumann.

Although mathematicians, physicists, and engineers were a majority, Wiener and von Neumann emphasized that their theories and models would be useful in economics and politics. This contributed to the humanistic aspects of cybernetics, to which among others anthropologists Margaret Mead (1901–1978) and Gregory Bateson (1904–1980) contributed.

At the end of 1951, there was a break between Wiener and McCulloch, as a result of which Wiener no longer participated in these conferences.

Participants in the 20th Macy Conference, when Wiener was no longer part of the group. Courtesy: Josiah Macy Jr. Foundation.

To deal with the problem of universals, McCulloch and Pitts used as a background idea the notion of geometric invariant proposed by the German mathematician Félix Klein (1849–1925) in 1872. For

Klein, we fix a geometry when we decide which movements of space will not modify the geometric properties that interest us. In the same way, McCulloch and Pitts thought that two images A and B could be considered by our brain as images of the same scene seen from different perspectives if there is a projective movement T of the space such that the image described by T(A) causes the state of excitation of the same set of neurons in the cerebral cortex as the neurons in that area that are excited by the image B. This is actually a very simplified version of their argument but gives the idea. In particular, they used two accepted neurological facts, namely, a neuron only admits the states of excitation and non-excitation, and the brain area with which we see is in the cortex. It is evident that the group G of transformations of ordinary space that we are considering induces a group of transformations $\tilde{G}$ on the cells of the cortex and that the previous definition of "images of the same scene" could have been made using this group $\tilde{G}$ directly. We have assumed that G is the group of projectivities, but we could, for example, restrict ourselves only to rigid movements, or consider a completely different group, such as the homeomorphisms of ordinary space. $\tilde{G}$ is necessarily a finite group (since there are only a finite number of neurons in the cortex), which makes it easier to define invariants associated with it. Assume that $\tilde{G}$ has N elements and let us consider an arbitrary functional f defined on the different scenes A (so that given the scene A, $f(A)$ is a real number). For each $g \in \tilde{G}$, we represent by g_A the scene that is obtained from applying the transformation g to the scene A. Then, $\alpha(A) = \frac{1}{N} \sum_{g \in \tilde{G}} f(g_A)$ represents an invariant, in the sense that if $B = g_A$ for some $g \in \tilde{G}$, then $\alpha(A) = \alpha(B)$. This is how universals can be defined.

Indeed, in their article, McCulloch and Pitts proposed two mechanisms that combined to produce an invariance of the objects to be recognized under two-dimensional scaling and translation. The first one submitted the image of the input stimulus to a group of uniform scaling transformations (dilations and contractions), and then computed the mean over all the transformations. The second centered the image on its center of gravity. In both cases, it was a question of making a system tolerant to certain transformations of the inputs that were somehow capable of correctly generalizing the concept they represent.

This article was one of the first in what was later known as pattern recognition, and in addition it showed a clear influence on Wiener, that was reflected in *Cybernetics*, through a chapter entitled Gestalt and universals. In it, Wiener deals with the problem of abstract pattern recognition, following the same approach proposed by McCulloch and Pitts. In addition, he takes the opportunity to speculate on the possible application of these ideas in medicine, for example, to help a blind person read a text or to construct prostheses.

Wiener's work and personal relationship with McCulloch and Pitts were already a fact at that time. Whereas McCulloch had met him in 1942 at a meeting that was preparatory for the Macy Conferences, Pitts was introduced to him by J. Lettvin (1920–2011) in 1943. Wiener was enchanted by Pitts's talent (with whom he discussed his proof of the ergodic theorem) and Pitts moved to Boston just to work with him.

J. von Neumann and Cybernetics

Cybernetics' interest in understanding the human mind and particularly the role of feedback in brain processes could not be ignored by those, such as Turing in England or von Neumann at Princeton, who were working on the theoretical and practical development of modern computers and automata. Indeed, their interest in the new ideas that were being developed in the Wiener–Rosenblueth environment quickly manifested itself through von Neumann, who from 1943 on decided to cultivate a professional and personal relationship with Wiener, even going so far as to found with him (in 1944) a scientific society, the "Teleological Society." The members, who would later be known under the name of "cybernetic group," would include in addition to von Neumann and Wiener the following names: Rosenblueth, McCulloch, Pitts, Bigelow, Aiken (whom we already know), H. H. Goldstine (1913–2004), from the University of Pennsylvania (who along with Aiken was the constructor of the Mark I and the ENIAC at Harvard), and R. Lorente de No (1902–1990), from the Rockefeller Institute. This group of people, which grew to include several prestigious names in the social sciences, such as G. Bateson (1904–1980) and M. Mead (1901–1978), would meet periodically under the auspices of the Macy Foundation to discuss cybernetics and its many applications.

John von Neumann
(1903–1957). Courtesy:
Los Alamos National
Laboratory, Public domain.

Von Neumann was impressed by McCulloch and Pitts' 1943 paper, where their neural model was explained, and also by Wiener, Rosenblueth, and Bigelow's article on teleology. Starting from the investigations of neurophysiologists, he proposed a computer prototype and set out to study automata theoretically. In particular, he provided an original solution to the functional stability of a complex automaton and demonstrated that the failure of some parts of an automaton — if it has a suitable structure — does not have to lead necessarily to its global failure. He also proved theoretically that it is possible to build an automaton that reproduces itself.

For the next five years, Wiener and von Neumann would visit each other regularly and have a lengthy discussion about the commonalities between cybernetics, computing, and brain physiology. At first, everything went very smoothly, but from 1946 on, things changed and the relationship became tense, reaching a state of absolute coldness. Still, both were for years the "soul" of the Macy conferences, where both were very comfortable and enjoyed discussing their different points of view.

The coldness that arose between them could have two distinct foci. The first of them was without a doubt the enormous distance between their political positions since the explosion of the atomic bomb in Hiroshima. While Wiener suffered the consequences of his progressive opinions through scientific isolation, caused by his manifest distrust of the military and even of the United States government, von Neumann was climbing up in the Administration, until in 1955 he was sworn in as a member of the Atomic Energy Commission. To make matters worse, von Neumann openly defended the implementation of a program to manufacture the super bomb.

But there was also a departure in the theoretical discussions. In November 1946, von Neumann had written a letter to Wiener in which he shared his well-founded doubts about the possible success of cybernetics in achieving an understanding of the human mind (or even that of any much more insignificant creature, such as an ant). In Heims' words:

> What bothered von Neumann was the state of experimental neurophysiology and its relation to theory. He said that the microscopic cytological techniques that were being used were as delicate as studying the circuits of an electronic computer by throwing stones at it or plugging the jet of a hose against them. On the other hand, formal-logical theories, interesting as they were in themselves, were so general that they did not provide specific predictions about the results of the experiments. According to von Neumann, the difficulties lie in the exceptional complexity of the human nervous system and, indeed, of any nervous system.

In his letter, von Neumann did not unload his criticism but proposed a way out, or at least what he considered could be a first step in the right direction:

> Now, the less-than-cellular organisms of the virus or bacteriophage type do possess the decisive traits of any living organism: They are self reproductive and they are able to orient themselves in an unorganized milieu, to move towards food, to appropriate it and to use it. Consequently, a "true" understanding of these organisms may be the first relevant step forward, and possibly the greatest step that may at all be required.

Indeed, von Neumann's letter included the outline of a long research program, the first step of which was to study

> (...) The physiology of viruses and bacteriophages and everything that is known about the gene-enzyme relationship.

Just a few years later, in 1952, A. Hershey and M. Chase carried out a series of experiments that served to confirm that the T2 phage transmits its genetic information in its DNA and, in 1953, J. Watson and F. Crick proposed the double helix model for DNA, which would lead to their being awarded the Nobel Prize in Medicine and Physiology in 1962.

Be that as it may, it is clear that Wiener and von Neumann did not reach an understanding, but on the contrary their positions with respect to their research interests were moving apart and, although both continued to attend the Macy conferences for several years, they left no doubt of their estrangement.

When it was von Neumann's turn to speak at the conference, Wiener would lie down to sleep, snoring loudly, or scribble very visibly, to make it clear that nothing von Neumann could say was of interest to him. For his part, when it was Wiener's turn to speak, von Neumann used to sit in the front row, and instead of listening he read the *New York Times*, making a lot of noise as he turned the pages.

Cybernetics and Feedback

As we know, during the Second World War, Wiener had worked on technologies for the improvement of automatic projectile guidance and had studied within that context the principle of feedback. According to this principle, the actual direction of the target to be destroyed with a guided projectile is compared to the direction that we had predicted according to a particular mechanism, and a negative correction is applied (negative in the sense that it "opposes" error). This description coincides with the one used by Wiener in his early work on "negative feedback," and it is not far from what many electronic devices do, such as thermostats. However, Wiener's view on feedback went far beyond the usual practice in handling feedback by physicists and electronics engineers. Wiener realized that the feedback principle is an intrinsic characteristic of living things, from the most basic single-celled algae to the most complex animals. According to this principle, living beings can be observed as a complex system that changes its actions in response to the environment.

Still, he was not the first person to associate the feedback with living things. In the 19th century, it could be considered that A. R. Wallace (1823–1913), a contemporary of Charles Darwin (1809–1882), proposed the theory that feedback over long periods of time is responsible for the evolution of species. The approach is interesting because it presents an analogy with the steam engine, which works according to the principles established in control theory. Interestingly, Wallace's claim seems to have been verified recently, at least at the protein level, since certain proteins can respond to specific mutations or disturbances in the chain they form, reconstructing it.

Wallace's exposition was more qualitative than quantitative, so it took several decades until the Italian mathematician V. Volterra

(1860–1940) formulated in 1931 a quantitative expression for the synergy between feedback and living beings. He established the balance between two populations of species in a fish farm, with a differential equations model that involved a feedback process. Volterra is one of the pioneers in the area known as mathematical biology and was undoubtedly one of Wiener's many inspirations.

Feedback control is the fundamental mechanism by which electrical and mechanical systems maintain their balance. Wiener explained that it is the same with biological systems and that it is through feedback that it is possible to achieve this equilibrium or homeostasis. For example, in higher life forms, the conditions in which an organism can survive can be narrow. Thus, a change in body temperature of as little as half a degree centigrade is generally a sign of illness. Wiener states in *Cybernetics* that the human body's homeostasis is maintained through the use of feedback control.

The First Physical Models of Feedback

The first artificially regulated system, a water clock, was invented by the Greek Ktesibios in the 3rd century BC. In his clock, water flowed from a fountain to a reservoir and then from the reservoir to the clock mechanism. Ktesibios's device used a cone-shaped float to measure the water level in the reservoir so that it never overflowed or stayed dry. It was probably the first purely artificial self-regulating mechanism that required no outside intervention between feedback and apparatus controls. Until the arrival of the steam engine, the only physical applications of feedback were different water clocks developed in various civilizations such as Greece, China, or Arabia.

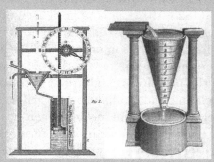

Two water clocks.

With the invention of the steam engine by J. Watt (1736–1819) in 1769, the study of teleological mechanisms (that is, mechanisms with a certain purpose) in machines with corrective feedback was also born. Indeed, one of the problems associated with the rotary steam engine is regulating its rotational speed. Taking advantage of a technology developed years earlier for mills, Watt developed a new form of speed regulation called a "governor." This governor consisted of a centrifugal valve to control the engine's speed and it would be Wallace's point of inspiration. Until that time, there was no serious feedback study. The developments mentioned earlier and other contemporaries were for practical use and mainly based on with intuition. There was still no attempt to generalize mathematically or to look for physical laws that would explain their operation in quantitative terms. It was not until a century later that, in 1868, Maxwell published "On the Governors," one of the first scientific works that discussed the principles of self-regulating systems, and the one that marks the end of the prehistory of control theory.

Indeed, as early as 1840, the mathematician and Greenwich Astronomer Royal, G. B. Airy (1801–1892), developed an automatic device that used feedback to point a telescope. This device was an automatic system for controlling the speed of rotation of the telescope to compensate for the effect produced by the rotation of the earth, allowing the astronomer to study certain celestial bodies in the same position for an extended time.

Although the applicability of Airy's system was quite limited, the merit of it was that he was the first to use differential equations in the analysis of a control problem. Furthermore, he also discovered that poor feedback loop design could cause strong oscillations in the system.

Since he was a child, Wiener had had a great interest in physiology and biological systems, so his contact with Rosenblueth from 1933 on and especially his work on antiaircraft batteries were very enriching for him. Precisely then, Wiener and Bigelow considered that very possibly the sensory and motor skills of the human being require the use of specific feedback processes. To confirm this hypothesis, and starting from the principle that the pathologies of an organ often shed light on its normal functioning, they asked Rosenblueth if there was any disease whose visible effect is that when a person tries to carry out a specific motor activity — for example taking a cup — his hand in a first attempt exceeds the limit of the cup, then corrects but without reaching it and quickly enters a loop of violent oscillations without achieving its objective. They asked this because this type of pathology frequently occurs in feedback systems that lose stability as a consequence of the fact that the regulatory effect of the feedback is excessive; if the error made in a first approximation is positive, the feedback changes the sign and makes it grow slightly. A second feedback changes the sign again and produces an increase in error, etc., until the system enters a process of violent oscillations, each time of greater magnitude. In control theory, this type of instability by excessive feedback is called "hunting," and is precisely the one described by Airy in his works. Obviously, Rosenblueth's immediate answer was that indeed such a disease exists in humans and is called cerebellar or volitional tremor because it appears as a consequence of damage to the cerebellum.

As a result of this observation, Wiener and Bigelow intensified their contact with Rosenblueth, and in 1943 the first joint article between them, "Behavior, purpose and teleology," appeared in the journal *Philosophy of Science*. As we already mentioned earlier, this paper is considered by some authors as the founding document of cybernetics. In it were proposed for the first time the hypotheses that all voluntary action implies a negative feedback process and that certain actions of the servomechanisms and of the animals can be studied in a unified way through the consideration of their purposes. It is worth commenting on some of the definitions that they introduced in the article. A mechanism M is said to be "intentional" if there is a coupling of the mechanism with some element E of its environment that causes in M a movement towards a certain state S_0 that maintains some kind of relationship with E. The same mechanism M is said to be teleological if it is intentional, and furthermore the movement of M towards the state S_0 is constantly controlled by the deviation of its current state S from the final state S_0. That is, the mechanism is teleological if a negative feedback process governs it.

A straightforward example of a teleological mechanism is the magnetized needle of a compass.

With these definitions (which obviously require the use of the second law of thermodynamics, since they need a time's arrow to exist), Wiener and his collaborators achieved something never seen before: They expelled all theological hypotheses to define "intentionality." In addition, they clearly raised the possibility that intelligent machines could be built sometime in the future.

With Rosenblueth, Wiener would later work on various physiological issues, eventually publishing several articles in the journals *Philosophy of Science*, *Archives of the Instituto de Cardiología Mexicana*, and *Journal of Comparative Physiology*. One of the conclusions they reached is that there are different forms of feedback in animal physiology. Specifically, we can distinguish between voluntary feedback (the one that has the greatest weight when we try to perform a conscious activity such as picking up an object or throwing a ball), postural feedback (which aims to maintain muscle tone), and homeostatic feedback (whose main motivation is to keep the body within reasonable parameters for its survival). For example, thanks to this last type of feedback, we keep our temperature around 37°, we take care that blood pressure is adequate, we breathe regularly, etc. One of the results that Wiener was able to demonstrate, and which is very important for cybernetics, is that, in general feedback, control of a mechanism that has several components cannot be carried out through a single feedback process, and sometimes two or even three simultaneous feedbacks are necessary. This result, which seems natural at first glance, was not easy to prove. Indeed, although Wiener dedicated a few paragraphs of the chapter on feedback and oscillation in *Cybernetics* to its demonstration, what he wrote was not easy to understand.

In the late 1940s, Wiener worked with J. Wiesner (1915–1994), who was then at the MIT Electronics Research Laboratory, to build an artifact (basically a train with photoelectric sensors) in which the different types of feedback that exist as well as their instabilities could be exhibited.

Wiener's ideas on physiology and feedback would be used from 1961 on to create human prostheses. In fact, Wiener himself was actively involved in designing the first artificial arm to be made in the United States, the Boston Arm, which was implanted in a person with considerable success in 1963. But this success, that should have brought him joy, was accompanied by a taste that was between bitter and bittersweet. On the one hand, it was clear that they had finally used his ideas with a laudable humanitarian goal: to make up, even if only partially, the loss of a limb to a war-mutilated man. On the other hand, the press was wrong to attribute the success to a doctor at the Liberty Mutual clinic rather than to Wiener. In effect, Bose remembered what happened because he was part of the team that launched the project:

> Then, in December 1963, the first press report on the project appeared in the *Saturday Review*. The article made passing references to Wiener as someone whose early theories were poetically appropriate, but named the young Harvard doctor, who also worked for the Liberty Mutual clinic, as the one who from a scientific point of view ... deserved applause for putting Wiener's theory at work.

And:

> I gave the newspaper to Wiener. He read the entire article and just tossed it aside. He did not speak. Not a word, not a criticism, nothing at all. (...)
>
> When the news reached the New York Times, full credit was given to this doctor, which was a hoax. He had not contributed anything to the project. He was just there and, on the last day, when we had a man with the electrodes on and things were already working, he appeared with the press, and the photos were taken (...)

A few years later, when the first marketable artificial arm became available, the patent went to the Liberty Mutual clinic, and what Wiener had wished would become a service to humanity, with which no one should profit, turned into a lucrative business for others, and they weren't even the ones who designed it.

Feedback and Oscillation

In Chapter 4, we included information that explained the basic elements of feedback. Now, we will briefly delve into certain particular cases that frequently appear in the study of wave filters. Specifically, we start from the hypothesis that the transfer function of the original filter and that of the feedback loop are both rational functions. The calculations that we are going to carry out are important because once understood, they allow the reading of the corresponding chapter on feedback and oscillation in *Cybernetics*, in which Wiener seems to forget that the reader does not know the ins and outs of linear systems theory. Although his arguments are essentially valid, the non-specialist reader loses direction and is inevitably left at the starting gate of one of the fundamental parts of the book. This also happens in other sections of *Cybernetics*, but an explanation of this one is of vital importance and, in addition, it is one of the most accesible sections.

The first thing we are going to do is to explain when the stability of a wave filter whose transfer function is a rational function occurs, and what it translates into in terms of the behavior of the filter. Then, we will see how to apply these results to certain feedback filters.

It is known that if

$$L(x)(t) = (x * h)(t) = \int\limits_{-\infty}^{+\infty} x(s)h(t-s)ds$$

is a convolution operator, its BIBO stability (that is, the property that the operator transforms, with continuity, bounded signals into bounded signals) is equivalent to

$$\int\limits_{-\infty}^{\infty} |h| < \infty.$$

Consequently, in the case of stability, the Fourier transform of the unit impulse response of the system exists and is a continuous function that vanishes at infinity. Therefore, the filter can be described in terms of its transfer function $H(w)$, that can be represented as $H(w) = R(iw)$, where

$$R(s) = \int\limits_{-\infty}^{\infty} h(t)e^{-st}dt$$

is the Laplace transform of h. On the other hand, the causality of the filter means that the convergence region of the Laplace transform of

h is precisely the half-plane of complex numbers whose real part is greater than the real part of the rightmost pole of the transfer function. If $H(w)$ is a rational function, the criterion becomes ensuring that all poles of $H(w)$ have negative real parts.

The previous arguments are somewhat intangible, since they do not give us a simple idea of what the stability of the system means. Thus, we give a different argument that in our opinion does shed light on this question. It allows us to perfectly visualize the response to the unit impulse of a system (which we will assume causal) whose transfer function is rational.

Assume, then, that our filter has a rational transfer function that for convenience we will write as

$$R(iw) = \frac{P(iw)}{Q(iw)} = \frac{b_m(iw)^m + b_{m-1}(iw)^{m-1} + \cdots + b_0}{a_n(iw)^n + a_{n-1}(iw)^{n-1} + \cdots + a_0}.$$

Then, it is a simple exercise to check that this function corresponds precisely to the transfer function of the system described by the linear differential equation of constant coefficients given by

$$a_0 y(t) + a_1 y'(t) + \cdots + a_n y^{(n)}(t) = b_0 x(t) + b_1 x'(t) + \cdots + b_m x^{(m)}(t).$$

(Take the Laplace transform on both sides of the equation, solve it, and evaluate at $s = iw$.) Also, if $R(iw)$ is the transfer function of a stable filter, it needs to vanish at infinity, so we know that, necessarily, $m < n$. Now, decomposing $R(s)$ as a sum of simple fractions (something we can do thanks to the fundamental theorem of algebra), we arrive at

$$R(s) = \sum_{k=1}^{N} \sum_{j=1}^{m_k} \frac{C_{kj}}{(s - \alpha_k)^j}$$

for certain complex numbers C_{kj}, where $\{\alpha_k\}_{k=1}^{N}$ are the poles of $R(s)$ and $\{m_k\}_{k=1}^{N}$ are their corresponding multiplicities. The interesting thing about this decomposition is that it allows us to easily find the inverse Laplace transform, which necessarily coincides with h, and is given by

$$h(t) = \sum_{k=1}^{N} \sum_{j=1}^{m_k} \frac{C_{kj}}{j!} t^j e^{\alpha_k t} u(t),$$

where $u(t)$ is Heaviside's unit step function. Now, if $\alpha = a + ib$ is a complex number, then

$$t^j e^{\alpha t} u(t) = t^j (\cos(bt) + i \sin(bt)) e^{at} u(t),$$

so each term of the previous expression is an oscillating function and the size of its oscillations either grows exponentially (if $Re(\alpha_k) > 0$), decreases exponentially (if $Re(\alpha_k) < 0$), or remains limited (if $Re(\alpha_k) = 0$) for $t \to +\infty$.

Thus, the system's response to the unit impulse is an absolutely integrable function if and only if $Re(\alpha_k) < 0$ for $k = 1, 2, \ldots, N$, which is what we wanted to prove.

Let us now assume that $R(iw)$ is the transfer function of the system. How do we find the poles of the rational function $R(z)$ to the right or to the left of the ordinate axis? We use a well-known result of complex variable functions that mathematicians call the "argument principle" and engineers call the "circumvallation property." According to this principle, if C is a closed curve in the plane and $f(z)$ is a complex variable function that is differentiable in a neighborhood of the curve C and at all points inside the curve, except perhaps a few points where it can have poles, and if we also assume that for all points $z \in C$, $f(z) \neq 0$, then,

$$\frac{1}{2\pi i} \int_C \frac{f'(z)}{f(z)} dz = N_0(f) - N_P(f),$$

where $N_0(f), N_P(f)$ denote, respectively, the number of zeros and poles of f inside the curve C (counted with multiplicity). It is assumed that the integral is taken by traveling the curve in the counterclockwise direction.

This result takes the name "principle of the argument" because the integral

$$\frac{1}{2\pi i} \int\limits_C \frac{f'(z)}{f(z)} dz$$

represents the number of turns that the curve $f(C)$ makes around the origin of coordinates (number of turns in the counterclockwise direction minus the number of turns in the clockwise direction).

Applying the principle of the argument to the function $R(s)$ on the curve C_M that is obtained by joining the interval $[-Mi, Mi]$ with the semicircle that joins the ends of this interval and is contained in the right half-plane $\Pi^+ = \{z \in C : Re(z) > 0\}$, and taking $M \to \infty$, we see that from a certain value of $M > 0$ on, the curve $R(C_M)$ gives the same number of turns around the origin (since this curve already encloses all the zeros and poles of $R(z)$ in the right half-plane Π^+).

Moreover, as $\lim_{|z|\to\infty} R(z) = 0$, we have that for sufficiently large M the part of the curve C_M corresponding to the semicircle does not contribute anything to the number of turns of C_M, so we could directly find the number of turns given by the image curve of the ordinate axis around the origin (adding the point of infinity, which is the origin). That value is necessarily equal to $N_0(R) - N_P(R)$, where $N_0(R), N_P(R)$, denote, respectively, the number of zeros and poles of R in the right half-plane Π^+. This implies that if we know the number of zeros of R in the right half-plane Π^+, then we can also find the number of poles in that domain.

The above arguments are especially useful when applied to the transfer function of a feedback system

$$R(s) = \frac{H(s)}{1 + \lambda G(s)H(s)}.$$

The poles of $R(s)$ are precisely the zeros of $\Lambda(s) = \frac{1}{\lambda} + G(s)H(s)$. Also, if we apply our previous argument to the curve $\Gamma = \{G(iw)H(iw) : w \in R\}$, we see that the number of turns that Γ does around the point $-\frac{1}{\lambda}$, which we denote by $n\left(\Gamma, -\frac{1}{\lambda}\right)$, satisfies

$$n\left(\Gamma, -\frac{1}{\lambda}\right) = N_0\left(\frac{1}{\lambda} + GH\right) - N_P\left(\frac{1}{\lambda} + GH\right)$$

$$= N_P\left(\frac{H}{1 + \lambda GH}\right) - N_P(GH).$$

Therefore, the feedback system is stable (that is, $N_P\left(\frac{H}{1+\lambda GH}\right) = 0$) if and only if

$$n\left(\Gamma, -\frac{1}{\lambda}\right) = -N_P(GH).$$

This is precisely H. Nyquist's (1889–1976) criterion for the stability of feedback systems, and is the one used by Wiener in the computations that he includes in the chapter in *Cybernetics* devoted to feedback and oscillation. With it at hand, Wiener computes several particularly significant examples, and among other things demonstrates the need for several simultaneous feedbacks to bring certain types of systems to stability.

Entropy and Information

Every expert in cybernetics knows that one of the fundamental concepts in this discipline is that of entropy. The main reason why this is so is that the behavior of living beings with respect to entropy is special. Entropy is a measure of disorder with the property that, when measured in a thermodynamically isolated system over time, it increases or at least never decreases. This property means that we can define a direction, or "arrow" for time, an essential instrument for the elaboration of all teleological concepts. On the other hand, living beings are not isolated systems but, on the contrary, are in constant energy exchange with their environment. Moreover, a property common to all living beings is that by organizing their immediate surroundings, they produce a local reduction of entropy in space and time. It turns out that this quality is not exclusive to living beings. Many machines, artificially created by man, also have the effect of locally reducing entropy. Wiener saw in this property an important point of connection between the living and inanimate mechanisms. The study of this connection is one of the fundamental themes of cybernetics.

But what exactly is entropy? There are essentially two ways to answer this question. The first, based on an in-depth study of heat engines, is how the concept arose historically. First, S. Carnot (1796–1832) in his *Reflections on the Motive Power of Fire and on Engines Suitable for Developing this Power* (published in 1824) showed that there is a limit to the maximum performance that can be achieved in a heat engine and that this limit is reflected by the expression $L = Q\left(1 - \frac{T_2}{T_1}\right)$, where L denotes the work done by the machine, Q the heat supplied to it from a thermal source at temperature T_1, and $T_2 < T_1$ is the temperature of the sink.

The important thing in this result is that the maximum performance does not depend on the good engineering that is used for the construction of the machine. In simple words, there is a fee that nature imposes on us in the form of heat which is lost in the environment for the production of work. Hence, Lord Kelvin (1824–1907) deduced that:

> There are no cyclical processes in which the heat absorbed from a heat source becomes entirely work.

And then, R. Clausius (1822–1888) stated that:

> Heat is not transferred from a low-temperature body to a high-temperature one unless this process is accompanied by a change that occurs elsewhere.

The two previous statements turn out to be mathematically equivalent. Furthermore, by properly manipulating Carnot's equation, it is possible to introduce a magnitude S (which we call entropy) that has the property that both the Kelvin and Clausius statements, if joined to the principle of conservation of energy, translate to the statement that the entropy value in a thermodynamically isolated system either remains constant — in which case we are dealing with a reversible process — or increases — in which case the process is irreversible. This is the Second Law of thermodynamics.

The second way to introduce entropy, which we consider easier to understand, is to use statistical mechanics. This procedure is the one used by Boltzmann and M. Planck (1858–1947), and we will explain it in detail in the following. We will begin by presenting a thought experiment, which should help us understand what happens to the molecules of gas when they are allowed to circulate freely in a container that keeps the gas isolated from the environment.

Let us suppose that we have an infinite plane in which we have placed coins and that these coins have been sharpened at the edge so much that they cannot stay on edge, so they forcefully fall with equal probability on heads or tails. Suppose that initially all the coins are heads. We move the plane, making the coins jump and fall randomly. What is the most likely coin distribution after a very large number of such tosses? That they all come out heads is highly unlikely. In general, any distribution is indicated by the number of coins that are heads since the rest are tails. Thus, there are $N + 1$ possible descriptions: no heads, one head, two heads, etc., till N heads. But not all of these distributions are equally likely because the particular set of coins that land heads can vary without the need to change its cardinality. So, if there are exactly k heads, this can occur in

$$\binom{N}{k} = \frac{N!}{k!(N-k)!} = \frac{1 \cdot 2 \cdots N}{(1 \cdot 2 \cdots k)(1 \cdot 2 \cdots (N-k))}$$

different ways. For example, there is only one way for all the coins to be heads, but there are N ways for exactly one of them to be heads, etc.

Bernoulli's law of large numbers tells us that with time we will have more and more mixed coin distributions, since the most probable case is that half of the coins are heads. Therefore, this type of distribution of the coins will appear with a much higher frequency than the highly ordered initial distribution of coins (all heads). This does not mean that the initial sort will never occur again, but it is quite improbable. The frequency with which it appears is so low that if we do the calculations, it may be necessary to wait longer than, for example, the current age of the universe. This situation occurs in fact for such low values of the number of coins as $N = 100$. Imagine what happens if we work with values that approximate the number of gas particles in a volume of one liter!

Suppose now that we have a system with N particles, which is thermodynamically isolated. This means that the system cannot exchange energy with the outside, so its energy will remain constant over time. Of course, each particle can vary its energy. For example,

when colliding with a slower one, a particle moving at high speed
will transmit a certain impulse to the slow particle that will make
it move faster, but simultaneously it will lose speed itself. Let's now
take into account the Planck quantization principle. It turns out that
each particle can change its energy level only in a limited number of
ways, and since the system contains a finite number of particles, it
follows that if the total energy is E and the possible energy levels are $E_1, E_2, \ldots, E_m$, then at each given instant there will be a
number n_1 of particles with an energy level E_1, n_2 particles with
an energy level E_2, etc., till n_m particles with an energy level E_m,
always satisfying the equalities $n_1 E_1 + n_2 E_2 + \cdots + n_m E_m = E$ and
$n_1 + n_2 + \cdots + n_m = N$. We are assuming an ideal situation, in which
there are no energies associated with the interaction between pairs
of particles, but the computations can also be done for that more
realistic situation.

Each determination of the list of numbers $(n_1, n_2, \ldots, n_m)$ defines
what in physics is called a microstate of the system. On the other
hand, the macrostate of the system will be determined by the
properties that we can observe at the macroscopic level, such as its
temperature, pressure, and volume. Therefore, it is natural that we
consider the possibility that two different microstates give rise to
the same macrostate of the system. Then, we can affirm that given
the physical conditions of the system, there will be an associated
microstate (that is, one that does not disturb the macrostate of the
system) that is more likely than all the others. When the system
reaches that microstate, we will say that it is in statistical equilibrium. Clearly, there can exist several different microstates that give
rise to statistical equilibrium.

It is clear that not all energy levels E_i are equally probable. This
implies that not all partitions $(n_1, n_2, \ldots, n_m)$ have the same probability to occur. Taking this into account, Boltzmann was able to
deduce a probability law for the different microstates; the probability of the microstate $(n_1, n_2, \ldots, n_m)$ is the number

$$P(n_1, n_2, \ldots, n_m) = \prod_{i=1}^{m} \frac{p_i^{n_i}}{n_i!},$$

where p_i denotes the probability that a particle is in the energy level
E_i. Once this probability has been calculated, the entropy of the

system associated with its microstate $(n_1, n_2, \ldots, n_m)$ is defined as

$$S(n_1, n_2, \ldots, n_m) = k \ln P(n_1, n_2, \ldots, n_m) = k \sum_{i=1}^{m} \ln \frac{p_i^{n_i}}{n_i!},$$

where $k = 1.38064852 \times 10^{-23}$ m^2 kg s^{-2} K^{-1} is Boltzman's constant.

From these data and by using simple optimization techniques, Boltzmann deduced the mathematical expression of a microstate corresponding to the statistical equilibrium of the system. This expression takes the form

$$n_i = \frac{N}{Z} p_i e^{-\beta E_i}, i = 1, 2, \ldots, m;$$

where

$$Z = \sum_{k=1}^{m} p_k e^{-\beta E_k}.$$

β is a value whose explicit expression we omit but about which we can say that it is directly related to the average energy of the particles in the system (especially important is the relationship $kT = 1/\beta$, where T is the temperature of the system).

If we let an isolated system evolve freely, it will tend towards statistical equilibrium, which means that its entropy will increase over time until it reaches its maximum value. This gradual increase in entropy is what justifies a multitude of physical phenomena such as, for example, that a gas tends to occupy the maximum volume that it is granted, that two gases that come into contact will mix, that hot bodies become cool (passing part of their heat energy to a cold body in contact with them), and that any heat engine needs a heat sink. Thermodynamic processes can therefore be divided into two classes: irreversible processes, which are those that lead to an increase in entropy, and reversible ones, which occur within the statistical equilibrium.

The fact that entropy grows over time causes substantial differences between past and future, providing an asymmetry in the temporal variable that, on the one hand, explains the irreversibility of certain phenomena and, on the other, distances us from classical Newtonian physics. This gave enormous value to the new statistical mechanics. Wiener would take advantage of these subtle differences between classical time and the new concept of asymmetric time (also called Bergsonian time) to introduce into cybernetics concepts such as "intentionality," "progress," or "capacity to learn" in animals and machines.

Although all these concepts were indeed exposed by Wiener in *Cybernetics*, who gave very precise details about their mathematical treatment, my personal opinion is that he expressed his point of view much better, and it is more easy to understand, in his *Human use of human beings.*

The idea of writing this essay was first suggested to him by the New York publisher A. A. Knopf (1892–1984) in early 1949, who proposed that he publish a text for the general public in which he explained the essentials of cybernetics and its social implications. Apparently, Wiener showed no interest at the time because he was overworked, lecturing here and there on the new discipline. However, the idea remained in his heart, so shortly after, when another editor (this time P. Brooks from Houghton Mifflin) proposed something similar, Wiener accepted immediately. After a first conflict with the publisher (Wiener sent a sample chapter that was not acceptable for a non-expert reader), he readjusted his view of the book and within a few months sent the manuscript.

This time, he managed to write a text whose content, fundamentally philosophical, was available to everyone. And the message he sent was important:

> The thesis of this book is that society can only be understood by studying the messages and communication facilities that it has and, furthermore, that, in the future, the messages sent between men and machines, between machines and men and between machines and machines, will play an increasingly preponderant role.

To carry out this study, it was, of course, essential to explain what we mean by message and how messages can be transmitted. To a certain extent, this is equivalent to providing a concept of information since the purpose of messages is to convey information. The definition and treatment that Wiener gives to this term throughout the book are based on the establishment of a certain parallelism between the spontaneous tendency of nature towards disorder and the natural tendency of messages to degrade. According to him:

> The orders by which we regulate our environment are a kind of information that we impart to it. Like any other kind of report, they are subject to deformations when passing from one entity to another. Generally, they arrive in a less coherent form and certainly no more coherent than the initial one. In communications and regulation, we always fight against the tendency of nature to degrade what is organized and destroy what makes sense, the same tendency for entropy to increase (...)
>
> We give the name of information to the content of what is the object of exchange with the external world while we adjust to it and make it accommodate us.

It is clear that on this basis the general tone and themes dealt with in the book are very philosophical in nature. It will suffice to name the titles of some of the chapters: "Progress and entropy," "Rigidity and learning. Two forms of communicative behavior," "The organization as a message," "Communications, secrecy and politics," and "Language, confusion and interference."

Information and Communication

Talking about communication thus implies talking about "quantity of information" and "messages." Wiener began to work on these concepts during his collaboration with Bigelow on prediction theory. While discovering the crucial role that feedback plays in human motor activity, Wiener and Bigelow had conjectured that the human mechanisms responsible for controlling motor activity could suffer from feedback problems and display the same "symptoms" as mechanical systems based on feedback. In addition, the question was raised whether the mechanisms for manipulating information in machines are similar to those present in living beings.

This open question gave rise to the thesis that communication problems and control problems are strongly linked. Thus, the analysis of a control problem implies the treatment of an underlying communication problem, which leads to dealing with the concept of "message" (either transmitted by electrical or nervous means). As messages can be transferred in noisy channels, their analysis involves modeling the noise and the relationship between the amount of information and entropy, and studying the meaning of these magnitudes in relation to the measure of organization or disorder of a system.

This idea of "quantity of information" is a key concept in the development of communication theory. It first appeared as part of R. A. Fisher's (1890–1962) research and later in the work of C. E. Shannon (1916–2001) and Wiener.

Claude E. Shannon (1916–2001). Courtesy: Mathematisches Forschungsinstitut Oberwolfach. Public domain.

Claude E. Shannon (1916–2001). He was born in Petoskey, Michigan. In 1932, he entered the University of Michigan, where he obtained a degree in mathematics and another one in electrical engineering, both in 1936. Then, he began his master's studies at MIT, where among other things he helped develop the Bush Differential Analyzer.

His master's thesis in 1937 was on logic circuits, Boolean algebras, and digital electronics. Called by one observer as "possibly the most important master's thesis of the 20th century," it was an important piece of work for the mathematical formulation of the foundations for the digital circuits that are part of all electronic devices.

During and after World War II, he worked for Bell Laboratories, and it is there in 1948 that he wrote his most famous work, "A Mathematical Theory of Communication," clearly influenced by Wiener. The theory developed has as its central point the concept of entropy and is known as the Shannon–Wiener information theory, or simply information theory. Subsequently, he published a book with W. Weaver, *The Mathematical Theory of Communication*, which expanded the content of the 1948 article, and made it more accessible.

In 1958, he returned to MIT where he was a professor until his retirement. He died in 2001, aged 84, after a long struggle with Alzheimer's.

Shannon developed in parallel to Wiener (and under his influence) the mathematical expression of the concept of information. Although R. Hartley (1888–1970) already spoke in 1928 of "information" as a measurable quantity, it was Shannon who, enlightened by Wiener's ideas on cybernetics, laid the foundations for the mathematical theory of information. This theory had as its central idea the quantification in some way of what is the information contained in a message.

Unlike Wiener, Shannon had a particular preference for the discrete rather than the analog. That is why, when elaborating an information theory (although he used the term communication theory), he started from a discrete probabilistic model. Let's see how he proceeded.

To begin, he defined a source of information as a process in which for each instant of time, a symbol is chosen at random from a certain (discrete) list of values that he called the alphabet. That is, for Shannon, setting an information source F is equivalent to setting a discrete stochastic process whose associated probability distribution is discrete (time is represented by instants $n = 0, 1, 2, \ldots$). Let us imagine that we have a source F that generates a succession of symbols that belong to the finite alphabet $A = \{x_1, x_2, \ldots, x_N\}$. If the symbols are emitted at each instant of time following a fixed probability distribution, and this is done independently of the values obtained in previous instants, we will say that we are dealing with a memoryless source of information. It is evident that such a source will be fully described from the probability distribution

$$P[X = x_i] = p_i \quad (i = 1, 2, 3, \ldots, N),$$

where X is the random variable that chooses at each instant of time the symbol x_i to be emitted by F. Having a memoryless source F is entirely equivalent to knowing the probability distribution of the associated random variable X. Thus, we can associate concepts such as information or entropy equally with the source F or with the random variable X.

To understand the concepts of Shannon's theory, we start by introducing the intuitive idea that for any message the following rule holds: the more unexpected, the more informative it is. For example, if today is Monday and someone tells us that tomorrow is Tuesday (a fact certain to happen), the information he is giving us is zero.

Furthermore, if we send two messages consecutively, but each of them is issued independently of the other, then we will have to admit that the information provided must be the sum of the information provided by each message.

These ideas have a straightforward mathematical expression: Self-information I of a value x of a discrete random variable X will verify the following:

- Self-information $I(x)$ is a value that only depends on the probability associated with the event $X = x$. That is, $I(x) = I(p)$, where $p = P[X = x]$. Furthermore, we assume that the dependency is continuous and that $I(x) = 0$ whenever $P[X = x] = 1$ (that is, if the value x represents a certain event).
- Let x_1, x_2 be two values of the sample space of X, which we denote by Ω_X. Then, $I(x_1) \geq I(x_2)$ if and only if $P[X = x_1] \leq P[X = x_2]$ (which means that a more unlikely event provides more information). In other words, $I = I(p)$ is a decreasing function of probability p.
- If we denote by $x_1 x_2$ the event "the source sends the symbol x_1 first and then the symbol x_2, and this is done independently," then $I(x_1 x_2) = I(x_1) + I(x_2)$. In other words, self-information I, interpreted as a function of probability p, verifies $I(p_1 p_2) = I(p_1) + I(p_2)$.

It is not difficult to prove that if $I = I(p)$ is a continuous decreasing function of $p \in (0, 1]$, which verifies $I(p_1 p_2) = I(p_1) + I(p_2)$, then

$$I(p) = K \log_2 \frac{1}{p} = -K \log_2 p$$

for a certain positive constant K.

A sketch of the proof would be as follows: First, it is proved by a double induction (one for the numerator and one for the denominator) that if r is a positive rational number, then $I(t^r) = rI(t)$ for every real number $0 < t < 1$. Next, we use continuity to assert the validity of the last equality for any positive real number r. We take $t = 1/2$ to get that $I(2^{-r}) = rI(1/2)$. Now, we force $p = 2^{-r}$, obtaining that $r = -\log_2 p$ and $I(p) = -K \log_2 p$, with $K = I(1/2)$, which is what we wanted to prove.

The above calculations justify Shannon's proposal for self-information:

$$I(x) = \log_2 \frac{1}{P[X = x]} = -\log_2 P[X = x],$$

when the event $X = x$ has a positive probability, and $I(x) = \infty$ if $P[X = x] = 0$.

The logarithm base determines the units of measurement of the information; thus, we have bits if the logarithm is in base 2, nats if the logarithm is Neperian, or hartleys if we take the decimal base. Note that the choice of bits as a unit of information is equivalent to forcing equality $K = I(1/2) = 1$. Therefore, 1 bit is the information provided by the occurrence of one of two equally likely outcomes.

Once we know the self-information associated with elementary events of the type $X = x$, we can define the entropy of the random variable X (equivalently, of the source F) as the mathematical expectation of the self-information

$$H(X) = \sum_{i=1}^{N} I(x_i)P[X = x_i] = \sum_{i=1}^{N} p_i I(p_i) = -\sum_{i=1}^{N} p_i \log_2 p_i,$$

where $p_i = P[X = x_i]$, $i = 1, 2, \ldots, N$.

This measure is continuous (a small change in the probability distribution produces a small change in entropy), symmetric (it does not depend on the order of how we take the probabilities), and maximum for the uniform distribution, which is the distribution $p_i = \frac{1}{N}$ for $i = 1, 2, \ldots, N$.

In 1957, the Russian mathematician A. I. Khinchin showed that this function is, except for a change of scale, the only one that satisfies the three properties just mentioned (in addition to another property related to the composition of random experiments, which is also very natural). Therefore, the formula above is the only reasonable option to define the entropy of a discrete random variable with a finite number of possible values. Obviously, what we have explained is only the first step to define entropy for a general source (which is not necessarily memoryless), but these computations show the essence of the concept, and the reader can be assured that a general concept is possible.

Although the definition of entropy has its origin in physics, the word has been used in many contexts of mathematics with very different meanings. For example, there is a concept of entropy for operators in Banach spaces that serves to measure their degree of compactness.

It was von Neumann who in an informal discussion they had on the subject suggested that Shannon use this name for his uncertainty measure. He stated two reasons: first, because a similar formula had been used in statistical mechanics under that name; second, he jokingly added:

> Nobody really knows what entropy is, so in a debate you will always have the upper hand.

The concepts of self-information and entropy are two of the fundamental pillars on which Shannon built his information theory. The other key concepts are channel (the source emits its messages through a communication channel, which can incorporate certain noise into the message, destroying a part of the information transmitted), encoding (which is the procedure by which we can reduce the negative effects produced by channel noise), and channel capacity, which is the highest information transmission rate allowed by a given channel, simply by choosing the source appropriately. Here, the transmission rate R is the average amount of information that passes through the channel per symbol, and is always less than or equal to the entropy of the source. The difference $H - R$ is called the equivocation.

With all these ingredients, Shannon demonstrated the following important result:

Shannon's Channel Theorem (1948). Let us consider a channel whose capacity is C and which is connected to an information source whose entropy is H. If $H \leq C$, then it is possible to encode the source so that the system outputs produce errors with an arbitrarily small frequency. On the other hand, if $H > C$, then for each $\varepsilon > 0$ there is a way to encode the source whose associated equivocation is less tan $H - C + \varepsilon$, but in no case can we encode the source to obtain an equivocation smaller than $H - C$.

One of the reasons this result is of fundamental significance for cybernetics is that, thanks to it, Cowan and Winograd were able to solve in 1963 (just a year before Wiener's death) a question of great importance for the understanding of how the brain works. Concretely, they observed that Shannon's Theorem serves among other things to demonstrate the existence of neural networks that work correctly despite having numerous neurons that either function incorrectly or do not work at all, a situation that we know is constantly occurring in the human brain where thousands of neurons die every day.

Wiener defined information himself, but he did so for the analog case. His ideas served as inspiration for Shannon, but it was the latter who hit the right key by presenting a theory adapted to new digital technologies, and that is why Shannon's approach became better known. However, it must be said that the interpretation given by Wiener is loaded with a much deeper physical and philosophical sense than anything we can find in Shannon's works. The importance of information to Wiener was such that he considered it one of the fundamental components of the universe. For example, in *Cybernetics*, we read:

> Information is information; neither matter nor energy.

For Wiener, matter, energy, and information are different physical phenomena, but none can exist without the other. The so-called "physical objects" (including living beings) are considered "persistent patterns of information" in a changing flow of matter and energy. Wiener believes that each physical process is a mixture of matter, energy, and information.

The interpretation goes further in the case of personal identity and human nature. Human beings are for Wiener "patterns of information persistent through changes in matter and energy." That is, despite the continuous exchanges of matter and energy in a person's body, the "complex organization" of a human (Shannon's information, which could be "encoded") is maintained to preserve the life, functionality, and personal identity. Wiener would later write the following observation in *The Human Use of Human Beings*, using a poetic tone:

> We are but whirlpools in a river of ever-flowing water. We are
> not stuff that abides, but patterns that perpetuate themselves.
>
> (...) The individuality of the body is that of a flame rather
> than that of a stone, of a form rather than of a bit of substance.

The Break Between Wiener and McCulloch

In 2005, a new biography [CoSi] of Wiener appeared in the US market whose title, *The Dark Hero of the Information Age. In Search of Norbert Wiener, Father of Cybernetics*, made it clear that the work, written by North American journalists F. Conway and J. Siegelman after eight long years of research, would focus on some obscure personal question. Although the other existing biographies (apart from Wiener's two autobiographical books) had dealt mainly with mathematics in the case of Masani [Ma] and politics in the case of Heims [He], the fundamental theme in this new biography was, in addition to insistently stating that Wiener's most important creation was cybernetics, the account of some unpleasant episodes in Wiener's family life. In particular, Margaret, Wiener's wife, is described as a sinister character who had among others the following "merits": being pro-Nazi (despite her husband's Jewish origin) and not hiding it even from her daughters; being a hyper-controlling person with her daughters, especially from their adolescence on; and in the sexual realm, a subject about which she was rather paranoid, even going so far as to make the accusation that they were trying to seduce their father (a ridiculous accusation, apparently motivated by the daughters having pierced their ears to wear earrings). In addition, according to the authors of that biography, the fault of the rupture between Wiener and McCulloch and consequently the decline of cybernetics lay with Margaret, who falsely accused "the boys"s that collaborated with Wiener and McCulloch of seducing Barbara in the middle of adolescence, during a season that she spent at the neurophysiologist's house. Margaret was aware of the falsehood of her accusation, but she felt uncontrollable jealousy towards McCulloch and his cyber group, from whom she wanted to separate Wiener, and she thought that her lies would achieve it. For his own, Wiener, who learned of this situation on one of his visits to Rosenblueth in Mexico, took advantage of the receipt of a letter from Pitts and Lettvin. Using pompous language, they requested him to work in their new laboratory at MIT. Wiener, unilaterally and without explanation, broke all contact with McCulloch's team. He separated himself from their group by sending the following telegram to Wiesner:

> Inpertinent letter received from Pitts and Lettvin. Please inform them that all relationship between me and your projects is permanently abolished. They are your problem. Wiener.

He then wrote a long letter to J. R. Killian Jr., then president of MIT. He accused McCulloch and his boys of trying to take over cybernetics and use MIT to obtain money and material resources to carry out a project for which they had not actually proven to be qualified. In addition, there were numerous personal comments in the letter. For example, he described McCulloch as "a colorful swashbuckling scientific hero whose appeal far outweighs his reliability" and accused Pitts of not having performed the tasks assigned to him as his assistant, and for which he collected his salary, and in addition having systematically failed in his attempts to finish his doctoral thesis. He then attached the letter he had received from Pitts and Lettvin so that Killian could see for himself its impertinence. Finally, he concluded by stating that he was not going to subject himself in any way to the whim of these "irresponsible young men" and that was why he was breaking off his relationship with them.

Although the letter from Pitts and Lettvin was not very respectful (among other things, it did not admit that Wiener was currently engaged in a project on the same level as theirs and, therefore, might not be available), it seems that Pitts and Lettvin's intention was not to disturb Wiener. They were just happy to share with him what they considered very good news. But Wiener was angry and could not see reality objectively, or was simply eager to have an excuse to break up.

In the above narrative, numerous details can be considered objective. For example, it is true that there was a rift between Wiener and McCulloch's group and that the separation was violent. Also, objective data are the letter from Pitts and Lettvin, the telegram from Wiener to Wiesner, and the letter to Killian. But the more intimate reasons for the exaggerated behavior are known to us only in a very indirect way.

Beyond this, there is a comment from Rosenblueth to Lettvin ten years after the telegram, and also the reaction of the daughters when they learned the story from the authors of the biography. Apparently, they weren't too surprised, and Peggy even went so far as to say that it was precisely the kind of "story" that could come from her mother's mind. Although Peggy saw in her mother's maneuver an attempt to keep Wiener under control, distancing him from other people who like McCulloch had increasing influence on him, Barbara — who was the object of the accusation — was also offended by the attitude of her father, who seemed to have doubted her morality and not raised the slightest question.

But can we consider these as objective actors in the brawl? Probably not, since it is evident that the relationship between Margaret and her daughters was bad and therefore their reaction could be not very objective. Further, with only the testimony of Rosenblueth, we do not have a guarantee of the veracity of the story.

Shortly after the biography by Conway and Siegelman appeared, it was reviewed in various media, and in general it seems that their version was accepted. On the other hand, they have been criticized for the excessive weight they give to cybernetics and because of the focus of the book, so close to the tabloid press. For example, in his review for the *Notices of the American Mathematical Society*, M. B. Marcus argues the following:

> I knew Wiener well enough to know that he was fiercely loyal and really very manly, despite his awkward appearance. He would have been devastated by the way Margaret is treated in this book and fighting mad. I could have lived without knowing all the dirt on my hero's wife. I'm glad Wiener never had to read this book.

On his part, F. Dyson, in his review for the *New York Review of Books*, offers an opinion very similar to ours, commenting that:

> [Rosenblueth's testimony alone] is not the kind of evidence that would convict a murderer in a trial.

Shortly after it appeared, the same magazine published the following thank you letter, in which Hope Franklin O'Neill, one of Wiener's living nieces, takes the opportunity to defend her aunt:

To the Editors:

As a niece and "surviving witness," I wish to express my appreciation to Professor Freeman Dyson for the bulk of his insightful review of the latest biography of my uncle, Norbert Wiener [*Dark Hero of the Information Age: In Search of Norbert Wiener, the Father of Cybernetics, NYR*, July 14]. Most particularly, I am touched by his penetrating summative comments regarding the book's destructive portrait of my Aunt Margaret: "And inevitably the reviewer wonders whether the story [offered by the daughters] is true. Margaret is now the accused and will never have a chance to answer... [having] left no friend behind to speak for her."

I cannot know all that my cousins Barbara and Peggy experienced, but I can speak up for my aunt, since I spent much time in the Wiener household over the years, and was Peggy's friend and classmate growing up, and visited my aunt in New Hampshire several times in her final decade. My uncle was also a frequent visitor at my parents' house in Belmont, where his mother Bertha lived for twenty years. My mother was his sister Constance and my father, mathematician Philip Franklin of MIT, was his longtime friend.

I remember both my aunt and uncle with true affection. I respect my aunt greatly, as a dedicated wife and mother — a good sport who did her best for both daughters growing up, while providing a secure and hospitable family nest for her mercurial and often demanding husband. She would host and entertain us cousins on many a week-long summer visit, feeding us royally, while simultaneously giving Norbert space for his scholarly activities and meeting his needs for a strict vegetarian diet. The most stable member of that gifted but unusual family, she certainly did not deserve to be painted as frigid or evilly scheming or crazy or quasi-Nazi — none of which she was.

In actuality, my aunt was an admirable and hard-working immigrant. Arriving in the US at age fourteen, and after surviving teenage privations in the West, she earned her way to graduate school at Harvard, raised her girls in places as far away as China, sturdily shouldered the not-easy burdens of her husband's growing fame, and tried, all the while, to give her daughters normal high school and college experiences and secure futures in the years beyond.

Dyson is right that this biography belongs to a fashionable contemporary genre "emphasizing the baring of family secrets and the exposure of human weaknesses" and it is no accident that the book's dust jacket is strikingly similar to that of *A Beautiful Mind*, the distinguished biography of manic-depressive mathematician John Nash. The reviewer is wrong, however, in crediting authors Conway and Siegelman with a "thorough job of historical research, interviewing most of the surviving witnesses...." In the first place, neither I nor Norbert's two other nieces were ever contacted by the authors although we were — and are — all very much alive. In the second place, throughout the book there is a paucity of comments by Norbert's really distinguished younger colleagues as opposed to ample quotes of the less stellar members of his back lot coterie.

It grieves me to see my relatives so misrepresented. I did not seek to be interviewed, but had I been, I would certainly have affirmed the fundamental decency of my aunt and uncle and queried the journalistic validity of relying so heavily on notoriously subjective personal interviews as primary sources. If tell-all interviews *are* to be used in this new age of scientific gossip, biographers should at least aim at presenting the broadest spectrum of personal and professional viewpoints. Only thus can they attain 360-degree credibility and avoid the sort of one-sided conclusions which assassinate the characters of their subjects.

Hope Franklin O'Neill
Pacific Palisades, California

It is evident that O'Neill's testimony is of great value, as she helps us soften Margaret's character, making her more human. But she can't be considered totally objective either. For example, in our opinion (and in all the reviews we know of), Conway and Siegelman did do a serious research on Wiener, but it is very likely that they were wrong in their approach. For them, the break with McCulloch and his team was a major blow against the later development of cybernetics, but, as M. B. Marcus so well points out,

> But here I think that they [Conway and Siegelman] are guilty of a misunderstanding of the nature of mathematical research that is prevalent among the nonmathematical public. That is, that mathematics, and perhaps scientific research in general, advances by the achievements of a very few extremely gifted individuals — people who are so deep that even their colleagues don't understand them. In this view Wiener's separation from Pitts and McCulloch doomed their effort in using the principles of cybernetics to explain the workings of the brain. Of course, collaboration with Wiener would have been helpful. But

McCulloch and Pitts were not dummies. They were tackling a problem that is still very far from a solution. There was enormous enthusiasm in the 1950s and 1960s for the revolutionary changes that would be brought about, not only by cybernetics, but also by artificial intelligence. Progress was made, and work is continuing. But the mysteries McCulloch and Pitts were trying to answer are amongst the deepest that exist.

In short, there is no doubt about the break itself, nor about the fact that it was motivated by personal issues that prevented dialogue and subsequent reconciliation. Still, it does not seem fair to place all the responsibility on a single person, who in addition was not the one who had to decide or act. Nor can it be said that the subsequent failure of cybernetics (if there has been such a thing) was fundamentally due to this breakdown.

God and Golem, or the Beginning of "Cyber-Ethics"

Wiener published *God and Golem, Inc.: A Comment on Certain Points where Cybernetics Impinges on Religion* in 1963. While previously his works of popularization had a marked scientific nature or expressed his worries on the possible impact of cybernetics on society, in this case, the work deals with a purely philosophical question, and the book can be considered the germ of a new subject: the ethics of new technologies, or, more specifically, the "cybernetic ethics."

Although the title seems to indicate it, it is not a book about religion or the relationship between religion and science, but about philosophy. Thus, religion is used metaphorically by establishing certain parallelism between it and cybernetics, which arises from the somewhat arbitrary identification between man and God and between machine and creation. In this way, following this allegory, he raises some common questions in religions that are translated into the study of learning machines, machines that reproduce themselves, and human–machine communication.

That there are machines capable of learning was already accepted at that time, since there was the example of a computer program capable of learning from its mistakes in the game of checkers (software created by A. Samuel, an IBM engineer, in 1959). For Wiener, the existence of these machines poses a problem similar to that presented in the *Book of Job*, where Satan (being created by God) plays a game with God for Job's soul. Could the creator lose the game played against his creation? In the case of Samuel's program, the answer was clearly "yes" if there was a clear criterion for "victory" where the learning phenomenon could occur. Although the program could be improved, and computing capacity was limited, Wiener predicted that in a few years it would be possible to win with the best of the masters in the checkers game. Thus, by relativizing the adjectives of "omnipotent" and "omniscient" as it seems to have been done in the *Book of Job*, an affirmative answer is given to the following question: Can any creator, even an unlimited one, play a relevant game with the creature he created?

From here, Wiener goes on to question what would happen if a machine programmed to learn caused the death of people. When the possibility exists that a machine learns and surpasses the creator, the relationships between them must be carefully reviewed. This is not the case of the Golem, the legend that gives the book its title, where according to tradition Rabbi Loew of Prague created a being of clay, to which he breathed life, to protect the ghetto from anti-Semitic attacks. The possibility that a machine (the "modern counterpart of the Golem") is more than a simple protector and that it can learn in diverse environments was already more than a reality. The danger of this relationship is illustrated in the following paragraph:

> No, the future offers very little hope for those who expect that our new mechanical slaves will offer us a world in which we may rest from thinking. Help us they may, but at the cost of supreme demands upon our honesty and our intelligence. The world of the future will be an ever more demanding struggle against the limitations of our intelligence, not a comfortable hammock in which we can lie down to be waited upon by our robot slaves.

The last point that Wiener raises is the possibility of machines that reproduce themselves. Since we can see man as a being created in the image and likeness of God, can there be the case of machines that create other machines? Can this creation be in the image and likeness of themselves? Wiener argues that this creation is possible and that even the creation of machines with certain variations by other machines can be allowed since nonlinear transducers can reproduce themselves. The argument includes the concept of "propagation of the race," which can be interpreted as a function where a living being creates another in its own image. Wiener shows that this "propagation of the race" can also happen in machines, occurring in machines whose representation is not only "pictorial" (with the same appearance) but also operational (with the possibility of performing the same tasks). It is interesting to note that von Neumann had already thought about this specific question, going so far as to affirm that there are automata that reproduce themselves.

Last Years

Wiener took advantage of the success of his publications to lecture in numerous places and to travel, which was one of the things he really liked. Thus, in 1951, he was a Fulbright lecturer in Paris (which he took advantage of to visit Spain); in 1955–1956, traveled to India, where he was received as a visiting professor in Calcutta; in 1960 and 1962, he spent a term at the University of Naples, Italy; and in 1964, at the time of his death, he held a visiting professor position in Holland. He also visited Russia (where cybernetics had suffered a full rejection in its early days, but in the wake of Stalin's death, the perspective changed and it became a highly respected discipline).

His visit to Madrid was short, with just enough time to give a lecture. His colleagues advised him to give the talk in French because, although he knew Spanish, it was very likely that his ideas were not to the liking of Franco's regime.

In addition to traveling, Wiener must have enjoyed the success of his "new science," as he liked to call cybernetics. In addition, awards were showered on him. The most important of all (see the chronology) was undoubtedly the National Medal of Science of the United States, an award that he received shortly before his death in 1964. It was personally presented to him by the then recently inaugurated President of the United States, L. B. Johnson (1908–1973).

Wiener was not in good health. In 1961, he had suffered a fall that caused his hip to break. Although he recovered pretty well from the mishap, he gained considerable weight. In 1963, he was diagnosed

Wiener receives the National Medal of Science from President Lyndon B. Johnson. From left to right: Jerome Wiesner (Presidential Science advisor), Norbert Wiener, John R. Pierce, Vannevar Bush, Cornelius B. van Niel, Lyndon B. Johnson, and Luis W. Alvarez. Courtesy: MIT Museum

with type 2 diabetes, and in addition it was detected that he had heart and hearing problems.

He died in Stockholm of a heart attack on March 18, 1964. He had gone to Sweden to give a lecture at the Academy of Sciences. While visiting the new Communications Laboratory of the Royal Institute of Technology, walking up some stairs, his heart failed. The autopsy revealed that he had suffered a pulmonary embolism, which caused the heart attack.

The news quickly reached the MIT, and the institute's flags were lowered at half-mast in honor of Wiener. That same night, a group of 21 people held a meeting to share memories of Wiener. They filled out their signatures on a sheet of paper that was sent to Margaret, with the simple claim "We love him." The signatures included many prominent MIT names, such as President Julius Stratton, and the founders (and directors) of several major MIT labs, such as the Rad Lab and the Lincoln Laboratory. And, there too were the signatures of Yuk Win Lee and Warren McCulloch.

Margaret, who had accompanied Wiener on his trip to Sweden, organized a funeral in Stockholm. He was cremated, and the ashes were shipped to the United States, where they were buried at Vittum Hill Cemetery in South Tamworth. A priest from the Episcopal Church conducted the ceremony. Then, several weeks later, there was a memorial in Belmont, run by the Unitarian Church that Margaret and her daughters regularly attended.

However, the decisive farewell, and the last to be formally organized, took place in the MIT chapel that Wiener used to attend to speak with Swami Sarvagatananda (1912–2009), a minister of the Boston Ramakrishna Vedanta Society who had been a friend of Wiener's for almost a decade, and with whom he frequently talked about reincarnation. Margaret visited Swami and asked him to give a memorial mass for Wiener. The celebration took place on June 2, 1964. It was attended by chaplains of all religions represented at MIT (Jews, Catholics, Protestants, and Hindus). The chapel was filled to such an extent that the Catholic priest had serious difficulties entering.

Margaret passed away in 1989, aged 95, after suffering from colon cancer for a long time. She is buried next to the ashes of her husband in Vittum Hill Cemetery.

Appendix: Wiener, Shannon, and the Rise of a Digital World

The classical Fourier analysis, as well as the extension achieved by Wiener with his Generalized Fourier Analysis, and some later developments, which allow these types of calculations to be carried out in the context of tempered distributions (also called generalized Schwartz functions), guarantee that it is feasible to represent "arbitrary" signals in the frequency domain and that such representation has various properties that make it a tool of enormous interest for physics and engineering. In particular, from the point of view of telecommunications, it is essential to understand this representation and to know how it affects the process of preserving and communicating all types of signals.

Here are two aspects that are of great importance: On the one hand, we want to be sure that when sampling physical signals of all kinds — ranging from sound to EKGs, EEGs, medical images, or even other signals from nature such as the vibrations produced by an earthquake, starlight, and cosmic background radiation — these samples allow us to fully identify the sampled signal and therefore reach irrefutable scientific conclusions. In this way, each specific scientific problem can be dealt with from the moment a device is available that allows samples to be taken with the appropriate speed and precision. On the other hand, the discrete nature of the digitized signal facilitates its storage and transmission between devices. These are precisely the two fundamental aspects that allowed the vertiginous rise of digital technology from the middle of the 20th century on.

When Shannon published "A mathematical theory of communication," he opted for the digital world. At the time, however, the predominant technology was analog. Wiener, who was already a well-known mathematician at the time, had introduced the language of wave filter theory, which is the appropriate language for working with signals that any telecommunications engineer studies today. But he had done so primarily in the context of analog signals. That is why the fundamental tool he used required mathematical theories typical of mathematical analysis, such as complex variable functions, series and the Fourier (and Laplace) transform, and probability theory. Many of these tools are no longer necessary when working with digital signals. Such signals arise when sampling an analog signal (discretization) — which is typically done by taking uniformly timed samples — and discretizing the possible values that one of these signals can take (quantization). When we do this, linear algebra replaces mathematical analysis, and probability theory can focus heavily on the discrete case. Both things undoubtedly represent an enormous simplification in the mathematical tools necessary to work with these signals. In addition, discretization and quantization allow all processes to be implemented on a computer, where they can be computed extremely quickly and accurately.

But neither Wiener nor Shannon had a digital computer in 1948. Of course, we have already observed that Wiener even considered the need to build one and that the project was essentially scrapped because it was thought to be a long-term project and immediate results were needed at that moment. However, Shannon realized that physical signals are necessarily band-limited and that such signals are wholly determined by uniform samples as long as the sample rate exceeds a certain limit that depends on the bandwidth of the signal. In addition, Shannon was also aware that there were already processes that should be considered by their very nature to be digital. An example was signals received by teletype.

Interestingly, finite-energy, band-limited signals had been mathematically characterized by Paley and Wiener. They are the elements of the space

$$X_W = \{x \in L^2(\mathbb{R}) : \operatorname{supp}(\hat{x}) \subseteq [-W, W]\},$$

whose study from the point of view of mathematical analysis was detailed by Wiener in his 1934 book *Fourier Transforms in the Complex Domain*. This is the same monograph in which Wiener explained the characterization in the frequency domain of causal filters, a result that he would later use in his work with Lee and Bigelow, where he introduced a significant part of the standard language of wave filters theory. These functions are, in fact, entire functions (holomorphic on the whole complex plane) that also have specific properties that limit the growth of their modulus as we move away from the origin of coordinates. Although it was already known that entire functions are very rigid, in the sense that knowing them over a sequence of points that converges to a point determines them completely, in the case of functions from the Paley–Wiener space, which is what the space X_W is now called, the stiffness is much higher. They are uniquely determined from uniformly spaced samples, as long as the space between consecutive samples is small enough.

Shannon–Whittaker–Kotelnikov's Sampling Theorem

As previously noted, all physical signals that we can measure and then preserve or send from one place to another are band-limited signals. That is, they are elements of the Paley–Wiener space X_W for some $W < \infty$, which will depend on the type of signal with which one works. In physical terms, this means that when a physical signal is measured in frequencies, there is a limit frequency above which there is no contribution of pure harmonics in the description of the signal. There are certain limits in the vibrational capacity of the particles. From a mathematical perspective, this is expressed by saying that the Fourier transform of the signal is identically zero outside a certain finite interval. That is, $\hat{x}(u) = 0$ whenever $|u| > W$, for a certain $W < \infty$. The smallest value $W \geq 0$ with this property is called the "signal bandwidth."

Statement and proof of the sampling theorem

Let us analyze what happens when the signal considered is periodic. In such a case, we will say that the signal has finite bandwidth if all its Fourier coefficients, except a few — a finite number of them — vanish. If the bandwidth is W, the signal will be a trigonometric polynomial

$$x(t) = \sum_{k=-N}^{N} c_k e^{\frac{2\pi i k t}{T}}, \text{ with } N/T \leq W,$$

and it will be determined by its value on a finite set of points.

In fact, if we take $P(z) = \sum_{k=0}^{2N} c_{k-N} z^k$, then we have that $x(t) = P(e^{\frac{2\pi i t}{T}}) e^{\frac{-2\pi i N t}{T}}$ and therefore the signal is completely determined from any set of $2N+1$ values $x(t_k)$ with $t_k \in [0, T)$, since these uniquely determine the polynomial. Thus, to recover the signal from uniform samples, it is enough to take the samples with a frequency greater than twice the bandwidth, $f_s = 1/h > 2W$, since in this case we will have that $2Wh < 1$ and, therefore , $2(N/T)h < 2Wh < 1$, which means that $t_{2N} = 2Nh < T$, and the uniform sampling requires at least $2N$ points from the interval $[0, T)$, which determines $P(z)$ unambiguously.

The sampling theorem states that this property holds even with aperiodic signals. Indeed, if the signal $x(t)$ is not periodic, but its Fourier transform vanishes for all frequencies $|u| > W$, then it is also possible to recover it from a discrete set of evenly spaced samples, although this time infinitely many values of the original signal will be required. The only condition necessary to ensure accurate signal recovery is that the sample rate is at least twice the signal's bandwidth.

There are several proofs of this theorem. The simplest is formalized for signals from the Paley–Wiener space, and except for the part dedicated to uniform convergence, it is the one that we can find in Shannon's original articles from 1948 and 1949: Suppose that $x(t) \in L^2(\mathbb{R})$ and $\hat{x}(u) = 0$ for $|u| > W$. Then, all the relevant information about the Fourier transform of the signal is located within the interval $[-W, W]$. Let us consider, in order to apply classical Fourier analysis, the signal $X(u)$ obtained by periodically extending with period $2W$ the restriction of $\hat{x}(u)$ to the interval $[-W, W]$. This signal lives in $L^2(0, 2W)$ and therefore coincides with its Fourier series:

$$X(u) = \sum_{k=-\infty}^{\infty} c_k(X) e^{\frac{2\pi i k u}{2W}} = \sum_{k=-\infty}^{\infty} c_k(X) e^{\frac{\pi i k u}{W}}.$$

Now, the computation of its Fourier coefficients provides a pleasant surprise, since it returns us the evaluation of $x(t)$ at the points of the form $k/(2W)$ with $k \in \mathbb{Z}$:

$$c_k(X) = \frac{1}{2W} \int_{-W}^{W} X(u) e^{-\frac{2\pi i k u}{2W}} du = \frac{1}{2W} \int_{-\infty}^{\infty} \hat{x}(u) e^{-\frac{2\pi i k u}{2W}} du$$

$$= \frac{1}{2W} x\left(\frac{-k}{2W}\right).$$

It follows that the Fourier transform of $x(t)$ is completely determined by the values that the signal takes at the points $k/(2W)$ with $k \in \mathbb{Z}$. Therefore, $x(t)$ is also determined by these values, which result from taking uniform samples of the signal with a sampling frequency equal to twice the bandwidth of the signal. This is just what we wanted to prove.

We can go a step further in this proof and compute an explicit formula that returns the signal $x(t)$ from its samples. To do this, it is enough to take into account that $\hat{x}(u) = X(u)\chi_{[-W,W]}(u)$, where $\chi_{[-W,W]}(u)$ is the characteristic function of the interval $[-W, W]$ — that is, it is equal to 1 for $u \in [-W, W]$ and 0 otherwise. Then, the Fourier series of $X(u)$ can be rewritten as

$$\hat{x}(u) = \sum_{k=-\infty}^{\infty} x\left(\frac{k}{2W}\right) e^{-\frac{2\pi i k u}{2W}} \frac{1}{2W} \chi_{[-W,W]}(u).$$

Taking the inverse Fourier transform on both sides of this identity, the sampling formula is obtained:

$$x(t) = \sum_{k=-\infty}^{\infty} x\left(\frac{k}{2W}\right) \operatorname{sinc}\left(2W\left(t - \frac{k}{2W}\right)\right),$$

where $\operatorname{sinc}(t) = \frac{\sin \pi t}{\pi t}$ is known as "sine cardinal" or "sampling function" (and the associated series is called "cardinal series").

Considering the argument we have just presented, the sampling formula must be interpreted in terms of convergence in the energy norm. However, a little trick (which appears already in Paley and Wiener's 1934 book) allows us to demonstrate the uniform (and absolute) convergence of the series.

Let us consider the errors of the series of absolute values:

$$E_N(t) = \sum_{|k|>N} \left| x\left(\frac{k}{2W}\right) \operatorname{sinc}\left(2W\left(t - \frac{k}{2W}\right)\right)\right|.$$

From the Cauchy–Schwarz inequality, we know that

$$E_N(t) \leq \left(\sum_{|k|>N} \left| x\left(\frac{k}{2W}\right)\right|^2\right)^{\frac{1}{2}} \left(\sum_{|k|>N} \left| \operatorname{sinc}\left(2W\left(t - \frac{k}{2W}\right)\right)\right|^2\right)^{\frac{1}{2}}.$$

Now, as $\{x\left(\frac{k}{2W}\right)\}$ are the Fourier coefficients of X, this sequence belongs to $\ell^2(\mathbb{Z})$ and

$$\left(\sum_{|k|>N}\left|x\left(\frac{k}{2W}\right)\right|^2\right)^{\frac{1}{2}} \to 0, \text{ for } N \to \infty.$$

The proof is complete if we can show that

$$\sup_{t\in\mathbb{R}}\left(\sum_{|k|>N}\left|\operatorname{sinc}\left(2W\left(t-\frac{k}{2W}\right)\right)\right|^2\right)^{\frac{1}{2}}$$

$$= \sup_{t\in\mathbb{R}}\left(\sum_{|k|>N}\left|\frac{\sin(\pi(-k+2Wt))}{\pi(-k+2Wt)}\right|^2\right)^{\frac{1}{2}} \le M$$

for a certain constant $M < \infty$ that does not depend on t or on N.

Now, if we consider the function

$$G(t) = \sum_{k=-\infty}^{\infty}\left|\frac{\sin(\pi(-k+2Wt))}{\pi(-k+2Wt)}\right|^2$$

and we do the change of variables $s = 2\pi Wt$, then $G(t) = H(s)$ where

$$H(s) = \sum_{k=-\infty}^{\infty}\left|\frac{\sin(-k\pi+s)}{-k\pi+s}\right|^2,$$

which is periodic with period π. Hence,

$$\sup_{t\in\mathbb{R}}G(t) = \sup_{s\in\mathbb{R}}H(s) = \sup_{s\in[-\pi/2,\pi/2]}H(s).$$

If $s \in [-\pi/2, \pi/2]$ and $k \ne 0$, then

$$|s - k\pi| \ge |k\pi| - |s| \ge |k\pi| - \pi/2 \ge |k|\pi/2,$$

so that

$$\left|\frac{\sin(-k\pi+s)}{-k\pi+s}\right|^2 \le \frac{1}{(|k|\pi/2)^2}$$

and

$$\sup_{s\in\mathbb{R}}H(s) \le 1+\sum_{|k|\ne 0}\left|\frac{\sin(-k\pi+s)}{-k\pi+s}\right|^2 \le 1+\sum_{k\ne 0}\frac{1}{(|k|\pi/2)^2} = K < \infty.$$

This ends the proof.

There are many variations of this result. For example, it can be proved that the sampling formula is satisfied by tempered distributions whose Fourier transform has compact support. The crucial step for that result is to show that if a tempered distribution is such that its Fourier transform vanishes outside of an interval, then the distribution is an ordinary function and it makes sense sampling it. Once this is clear, the proof is based on a very elegant calculation that requires a good understanding of the properties of the pulse train with period T,

$$\mathrm{III}_T\{\phi\} = \sum_{n=-\infty}^{\infty} \phi(nT).$$

It is easy to check that the following identities are satisfied (in a generalized sense):

$$x_s(t) = \sum_{n=-\infty}^{\infty} x(nT)\delta(t - nT) = \sum_{n=-\infty}^{\infty} x(t)\delta(t - nT)$$

$$= x(t) \sum_{n=-\infty}^{\infty} \delta(t - nT) = x(t)\mathrm{III}_T(t).$$

It follows that

$$\hat{x}_s(\xi) = (\hat{x} * \widehat{\mathrm{III}}_T)(\xi)$$

$$= \left(\hat{x} * \frac{1}{T} \sum_{n=-\infty}^{\infty} \delta\left(\cdot - \frac{1}{T}n\right)\right)(\xi)$$

$$= \frac{1}{T} \sum_{n=-\infty}^{\infty} \left(\hat{x} * \delta\left(\cdot - \frac{1}{T}n\right)\right)(\xi)$$

$$= \frac{1}{T} \sum_{n=-\infty}^{\infty} \hat{x}\left(\xi - \frac{1}{T}n\right).$$

Obviously, the above expression implies that $\hat{x}_s$ is $\frac{1}{T}$-periodic. Also, if $\frac{1}{T} \geq 2W$, then $[-W, W] \subset \left[-\frac{1}{T}, \frac{1}{T}\right]$. Furthermore, for each $\xi \in \mathbb{R}$ there is a unique $n \in \mathbb{Z}$ such that $\xi - \frac{1}{T}n \in \left[-\frac{1}{T}, \frac{1}{T}\right)$. It follows that in the sum

$$\hat{x}_s(\xi) = \frac{1}{T} \sum_{n=-\infty}^{\infty} \hat{x}\left(\xi - \frac{1}{T}n\right)$$

all members except one vanish since $\mathbf{supp}(\hat{x}) \subseteq [-W, W]$. Thus, $\hat{x}_s$ coincides with the $\frac{1}{T}$ -periodic extension of $\frac{1}{T}\hat{x}$, which implies that

$$\hat{x}(\xi) = T\hat{x}_s(\xi)\chi_{[-W,W]}(\xi), \quad \text{for all } \xi \in \mathbb{R}.$$

Hence,

$$x(t) = T\mathcal{F}^{-1}\left(\hat{x}_s(\xi)\chi_{[-W,W]}(\xi)\right)(t), \text{ all } t \in \mathbb{R}.$$

It follows that x is fully determined from x_s and henceforth from the samples $\{x(nT)\}_{n\in\mathbb{Z}}$. Moreover, if $T = \frac{1}{2W}$, then

$$\hat{x}(\xi) = \frac{\pi}{b}\hat{x}_s(\xi)\chi_{[-W,W]}$$

$$= \frac{1}{2W} \sum_{n=\infty}^{\infty} x\left(\frac{n}{2W}\right) \overline{\delta\left(\cdot - \frac{n}{2W}\right)}(\xi)\chi_{[-W,W]}(\xi)$$

$$= \frac{1}{2W} \sum_{n=-\infty}^{\infty} x\left(\frac{n}{2W}\right) \exp\left(-\frac{n}{2W}i\xi\right) \chi_{[-W,W]}(\xi)$$

for all $\xi \in \mathbb{R}$. Thus,

$$x(t) = \sum_{n=-\infty}^{\infty} x\left(\frac{n}{2W}\right) \mathcal{F}^{-1}\left(\frac{1}{2W}\exp\left(-\frac{n}{2W}i\xi\right) \chi_{[-W,W]}(\xi)\right)(t)$$

$$= \sum_{n=-\infty}^{\infty} x\left(\frac{n}{2W}\right) \text{sinc}\left(2W\left(t - \frac{n}{2W}\right)\right),$$

which is what we wanted to prove.

If we want to apply the sampling theorem in physics and engineering, this distributional version is essential since band-limited tempered distributions represent a case that is general enough to contain as particular cases those situations in which the theorem is to be used. Indeed, the proof of the sampling formula that we have just introduced requires that the function $x(t)$ be of finite energy, which is not the case with functions like $\sin 2\pi kt$, $\cos 2\pi kt$, which are so important in applications. However, these functions, seen as generalized functions, are tempered distributions with compact-support Fourier transform, and therefore the sampling theorem in its distributional version can be applied to them. Another example to which we can apply the same argument is the case of band-limited quasi-periodic functions.

Thus, the distributional proof of the sampling theorem is an interesting example of how generalization is sometimes necessary in mathematics. Hadamard, in a famous quote, claimed that "the shortest path between two truths in the real domain passes through the complex domain," alluding to how extremely useful some results of complex variables, such as the residue theorem, can be to demonstrate certain properties of functions of a real variable. Similarly, the distributional proof of the sampling theorem is a clear example of the use of generalized functions to prove an important property for ordinary functions.

When Shannon published his results, the theory of distributions was not yet developed, because despite the fact that Wiener had already introduced the concept of generalized function, the Fourier transform for tempered distributions appeared in the work of L. Schwartz (1915–2002) in 1950. However, Shannon's view of the applicability of the sampling theorem was certainly radical. It consisted of simultaneously limiting the bandwidth W of the signal and the time interval $[0, T]$ in which the signal is defined, stating that

> (...) any function limited to the bandwidth W and the time interval T can be specified by giving $2WT$ numbers.

After all, it is clear that every "real-world" signal must have a beginning and an end. However, from a purely mathematical perspective, these hypotheses are contradictory because the uncertainty principle, in its qualitative version, tells us that if we limit the bandwidth of a signal, its support in the time domain cannot be compact, and also vice versa: If the signal is limited in time, it cannot be band-limited. However, as D. Slepian (1923–2007) would claim later, there is no real difference between "being zero" and "being extremely small" for an engineer or a physicist. Shannon argued that our interest will always focus on a finite time interval. Now, assume that we restrict ourselves to a classical interpretation of the sampling theorem, and we take a band-limited signal with bandwidth W. In that case, although we are only interested in recovering it on the time interval $[0, T]$, it will not be enough to take evenly spaced samples in that interval because the cardinal series could contain infinitely many non-null terms. This is precisely what is observed in the following figure:

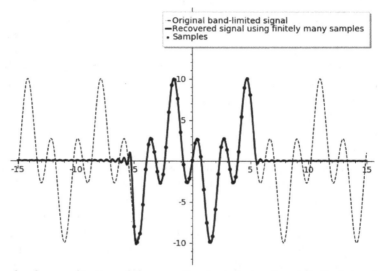

Example of an application of the sampling formula to a band-limited signal using only a finite number of samples. As can be seen, it is only retrieved with fidelity (but never with absolute precision) near the interval where the sampling was performed.

However, what is certain is that the approximation obtained in the interval where the samples have been taken is reasonably good (but not exact), and that as soon as we move away from it, the signal begins to take values very close to zero. Still, this situation was unpleasant if Shannon was trying to develop a mathematically correct yet elegant theory.

To give the character of a "theorem" to his claim that "$2WT$ samples are sufficient," Shannon employed a brilliant trick. He adapted to his interests the definition of "being limited in time" (which in a classical context means that $x(t) = 0$ whenever $t \notin [0, T]$), by substituting it with the following definition, adapted to band-limited signals: we say that the band-limited function $x(t)$ with bandwidth W is time-limited to the interval $[0, T]$ if $x\left(\frac{k}{2W}\right) = 0$ when $\frac{k}{2W} \notin [0, T]$. That is, a signal with bandwidth W will be said to also have limited time T if the only non-null samples that appear in the associated cardinal series are precisely those that fall within the interval $[0, T]$. Of course, this definition can obviously be adapted to any time interval, and it is particularly interesting to use the interval $[-T/2, T/2]$. With this interpretation, it is clear that the result formulated by Shannon is true, with no need to use distribution theory or other sophisticated mathematics. In this way, Shannon identified the signals with bandwidth W and time-limited to an interval of length T with ordinary space of dimension $2WT$, a result that is known as "Shannon's dimensionality theorem." And it is precisely this geometric interpretation that he later used to study the transmission of signals in the presence of noise.

Shannon's proposal was audacious. Another equivalent solution would be to adapt the statement of the theorem by claiming that every band-limited signal that vanishes on the Nyquist samples that do not belong to a given interval of length T is fully determined by $2WT$ samples in that interval. But neither Shannon's solution nor this one is completely satisfying, since one wants to deal with arbitrary band-limited signals, and a deeper understanding of the role that the uncertainty principle has on real-world signals is necessary. Slepian was the person who faced this question in the 1970s, and he found a clever useful solution. The key point is to distinguish real-world signals from their mathematical models, the ordinary functions. Real-world signals are those we get taking measurements with physical devices, and this imposes a certain precision threshold $\varepsilon > 0$. Concretely, we can assume that two ordinary functions $s_1(t)$, $s_2(t)$ are really indistinguishable at level ε if the energy of their difference satisfies that

$$E(s_1 - s_2) = \int_{-\infty}^{\infty} |s_1(t) - s_2(t)|^2 dt \leq \varepsilon.$$

Then, a function $x(t)$ is time-limited to the interval $[-T/2, T/2]$ at level ε if it is really indistinguishable at level ε from the function

$$y(t) = x(t)\chi_{[0,T]}(t) = \begin{cases} x(t), & \text{if } t \in [-T/2, T/2] \\ 0, & \text{otherwise.} \end{cases}$$

Similarly, $x(t)$ is band-limited to the interval $[-W, W]$ at level ε if it is really indistinguishable at level ε from a function which is band-limited to $[-W, W]$ in the classical sense. That is, $x(t)$ is really indistinguishable from $y(t)$ at level ε and $\mathbf{supp}(\hat{y}) \subseteq [-W, W]$. It is now clear that, given any $\varepsilon > 0$, every real-world signal is simultaneously time-limited and band-limited at level ε for some $T, W > 0$. Thus, to deal with arbitrary real-world signals, is sufficient since real-world signals have finite energy.

To recover Shannon's dimensionality theorem, it is necessary to add a new definition. We say that a set of functions $\mathcal{X}$ has approximate dimension N at level ε in the interval $[-T/2, T/2]$ if there is a fixed set of $N = N(T, \varepsilon)$ functions $\{\phi_1, \ldots, \phi_N\}$ such that every element of $\mathcal{X}$ is really indistinguishable at level ε during that interval from some function of the space spanned by $\{\phi_1, \ldots, \phi_N\}$, and there is no set of $N - 1$ functions that satisfy the same property. The dimensionality theorem was proven by Slepian with the following statement: Let $\mathcal{X}_\varepsilon$ be the set of functions that are band-limited to $[-W, W]$ and time-limited to $[-T/2, T/2]$ at level $\varepsilon > 0$. Let $N(T, W, \varepsilon, \epsilon)$ be the approximate dimension of $\mathcal{X}_\varepsilon$ at level ϵ during the interval $[-T/2, T/2]$. Then, for every $\epsilon > \varepsilon$,

$$\lim_{T \to \infty} \frac{N(W, T, \varepsilon, \epsilon)}{T} = 2W \text{ and } \lim_{T \to \infty} \frac{N(W, T, \varepsilon, \epsilon)}{W} = 2T.$$

Moreover, for every $\eta > 0$, there is a number $c(\eta, \varepsilon, \epsilon)$ such that, if $WT \geq c(\eta, \varepsilon, \epsilon)$, then

$$1 - \eta \leq \frac{N(W, T, \varepsilon, \epsilon)}{2WT} \leq 1 + \eta.$$

In particular, this means that in the limit, we only need $2WT$ samples to fully identify a signal that belongs to $\mathcal{X}_\varepsilon$, and it is WT that is the important magnitude we must consider when dealing with the transmission of signals. In particular, when using the sampling formula, if we have that $n = 2WT >> 1$, then n samples are enough to essentially determine any signal which is time-limited and band-limited (with time size T and bandwidth W) at level ε. In other words, the results can be summarized by saying that a signal of bandwidth W can be specified by $2WT$ samples each T seconds, when T tends to infinity. Presented in this form, Shannon's dimensionality theorem is a result about limits, and it seems that this was indeed the idea that Shannon had in mind when working with concepts such as the capacity of a channel.

On the meaning and history of the sampling theorem

Let's think for a moment what the sampling formula means. It is something revolutionary. Thanks to it, we know that any one-dimensional band-limited signal is entirely determined from its samples, as long as the sampling frequency exceeds a threshold that only depends on its bandwidth. As it turns out that all one-dimensional physical signals that we can measure are necessarily band-limited, the formula is universal. It can be applied to everything that is measured. In particular, the sampling formula tells us that the world can be discretized in time without losing information.

As expected, the impact this discovery had on the scientific community was immense. On the one hand, there are technological applications. If we want to faithfully preserve the information contained in electrical signals, audio, and even in images and videos — because the theorem admits multidimensional versions — it will be enough to choose a sampling frequency that is twice the signal's bandwidth. Each application requires a specific sampling frequency, which depends on the type of signals with which we are going to work. Technological devices designed for applications should take into account how often it is necessary to take measurements. Our physical knowledge of the signals will provide us with the information required to decide the appropriate sampling frequency in each case. For example, we know that the human ear only picks up sounds between 20 and 20,000 Hz, so the bandwidth of sound signals should have at most this last magnitude. On the other hand, the sound related to language moves at much lower frequencies, in the band between 200 and 3,400 Hz, information that is useful for the technology of standard telephony.

The research subjects that require devices to collect experimental measurements are very diverse. And it is extremely important that the measures taken by them for subsequent analysis take into account the dictates of the sampling theorem. Only then will scientists ensure that they have the necessary information and that nothing escapes between samples.

The sampling theorem was formulated for the first time in 1915 by the British mathematician E. T. Whittaker (1873–1956), who was interested in studying the interpolation properties of the cardinal sine function. It was independently proved five years later by the Japanese Kinnosuke Ogura. In 1928, the Swedish-American physicist and engineer Harry Nyquist (1889–1976) formulated the result without actually demonstrating it. In his honor, Shannon christened the sampling frequency $f_s = 2W$, the lowest from which we can recover the original signal, the "Nyquist frequency."

Sampling Above and Below the Nyquist Frequency

It is possible to apply the sampling formula — which allows the conversion of a signal in continuous time to a signal in discrete time — if we assume a bandwidth $\tilde{W}$ above the true bandwidth of the signal. It suffices to observe that if $\tilde{W} > W$, then $[-W, W] \subseteq [-\tilde{W}, \tilde{W}]$, and therefore the Fourier transform of the signal also vanishes identically outside of the interval $[-\tilde{W}, \tilde{W}]$.

For this reason, all the computations made in the proof of the sampling theorem remain valid, and the formula that returns the signal from its samples can be applied if we oversample, that is, if the sampling frequency exceeds the Nyquist frequency, $f_s = 2W$. In fact, doing so has the advantage of reducing the effect of rounding errors, resulting in a more stable version of the original theorem.

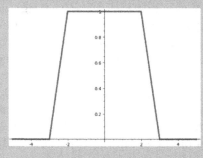

Moreover, if instead of multiplying by the characteristic function of $[-\tilde{W}, \tilde{W}]$, we multiply on both sides of the Fourier series by the trapezoid of the figure, which vanishes outside of $[-\tilde{W}, \tilde{W}]$ and is equal to 1 in $[-W, W]$, we will obtain the expression

$$x(t) = 4\pi^2 \sum_{n=-\infty}^{\infty} x\left(\frac{n}{2\tilde{W}}\right)$$
$$\times \frac{\cos\left(2\pi W\left(t - \frac{n}{2\tilde{W}}\right)\right) - \cos\left(2\pi\tilde{W}\left(t - \frac{n}{2\tilde{W}}\right)\right)}{\tilde{W}(\tilde{W} - W)\left(t - \frac{n}{2\tilde{W}}\right)^2},$$

for the inverse Fourier transform which, thanks to the square in the denominator, is much more stable.

It should be noted that the sampling formula fails if it is applied with a sampling frequency smaller than the Nyquist frequency, as it is unable to accurately detect the high frequencies of the original signal. In such a case, the recovered signal is what we call an "alias" of the original signal and is accurate only on a smaller bandwidth than the actual signal bandwidth. Moreover, if we fix an interval $[0, T]$, any physical signal defined on $[0, T]$ can be well approximated by signals that are band-limited and time-limited to $[0, T]$, just by taking the bandwidth $\tilde{W}$ large enough.

In the Soviet Union, the attribution of the Sampling Theorem fell to the engineer V. Kotélnikov (1908–2005), who proved it independently in 1933. Americans attributed it to Shannon, who introduced it in 1948 in engineering with "A Mathematical Theory of Communication." Shannon was apparently unaware of Kotélnikov's work, which led him to claim that:

> [This theorem] has been given previously in other forms by mathematicians [and he includes a citation to J. M. Whittaker's book of 1935 on interpolation theory] but in spite of its evident importance seems not to have appeared explicitly in the literature of communication theory.

However, the article published by Kotélnikov in 1933, from the title — "On the channel capacity of the 'ether' and of the wire in the electrical communications" — already made it clear that he was fully aware of the importance of the theorem for communications engineering. The first paragraphs also show a very evident determination, directed towards the same goals that Shannon would achieve independently a few years later:

> In both wired and radio engineering, any transmission requires the use of not simply a single frequency, but a whole range of frequencies. As a result, only a limited number of radio stations (broadcasting different programmes) can operate simultaneously. Neither is it possible to convey more than a given number of channels at any one time over a pair of wires, since the frequency band of one channel may not overlap that of another: such an overlap would lead to mutual interference.
>
> In order to extend the capacity of the "airwaves" or a cable (something that would be of enormous practical importance, particularly in connection with rapid developments in radio engineering, television transmission, etc.), it is necessary either to reduce in some way the range of frequencies needed for a given transmission (without adversely affecting its quality), or to devise some way of separating channels whose frequencies overlap — perhaps even employing a method based not on frequency, as has been the case until now, but by some other means?
>
> At the present time no technique along these lines permits, even theoretically, the capacity of the "airwaves" or a cable to be increased any further than that corresponding to "single side-band" transmission. So the question arises: is it possible, in general, to do this? Or will all attempts be tantamount to efforts to build a perpetual motion machine?
>
> This question is currently very pressing in radio engineering, since each year sees an increase in the "crowding of the airwaves." It is particularly important to investigate it now in connection with the planning of scientific research, since in order to plan, it is important to know what is possible, and what is completely impossible, in order to direct efforts in the required manner.
>
> In the present paper this question is investigated, and it is demonstrated that for television, and for the transmission of images with a full range of half tones, and also for telephony, there exists a fully determined minimum necessary frequency band, which cannot be reduced by any means without adversely affecting quality or speed of transmission. It is further demonstrated that for such transmissions it is impossible to increase

either wireless or wired capacity by any means not based on frequency bands — or, indeed, any other method (except, of course, by the use of directional antennas for separate channels). The maximum possible capacity for these transmissions can be achieved through "single sideband transmission," something fully achievable in principle at the present time.

For transmissions such as telegraphy, or for the transmission of images or television pictures without half tones, where the source may not change continuously, but is limited to specific, predetermined values, it is demonstrated that the required bandwidth can be reduced as much as desired, without adversely affecting the quality or speed of transmission, but at the expense of increasing the power and the complexity of the equipment. One such method of bandwidth reduction is outlined in the present paper, with a discussion of the necessary power increase.

There is thus no theoretical limit to the capacity of either the "airwaves" or a cable for transmissions of this kind; it is simply a matter of technical implementation.

It is fair, however, to note that Kotélnikov restricted his proof to functions that are absolutely integrable on the real line — that is, he assumed that the functions to which the theorem applies were elements of $L^1(\mathbb{R})$ — which, moreover, verify Dirichlet's conditions for the convergence of the Fourier transform (they have a finite number of maxima, minima, and discontinuities in any given finite interval). Thus, Shannon's proof, when complemented with the argument of uniform convergence, is much more general.

Actually, the list of names linked to the sampling theorem is pervasive, as many researchers proved it without being aware that they were working with a known result. Currently, it is known as the "Shannon–Whittaker–Kotélnikov sampling theorem," listing the authors in order of who contributed the most to its creation by publishing it and finding a way to apply it in various contexts. We know that long before Shannon used it in his 1949 article, Wiener was also familiar with the sampling theorem, since he formulates it explicitly in his report *Interpolation, extrapolation and smoothing of stationary time series*. Here, he also quotes his book with Paley (from 1934), which in turn cites the works of J. M. Whittaker (son of E. T. Whittaker) from 1927 and 1928, where the cardinal series is studied and the trick that demonstrates the uniform convergence of the series is shown. Furthermore, Wiener uses the cardinal series to demonstrate an important generalization of Liouville's theorem, due to G. Pólya (1887–1985), which guarantees that the entire functions that are bounded on the integers and have the additional property that

$$\overline{\lim}_{r\to\infty}\frac{1}{r}\max_{\theta\in[0,2\pi]}\log|f(re^{i\theta})| = 0$$

are necessarily constant.

It is worth quoting here a few phrases by H. D. Lüke (1935–2005), who in an article published in 1999 commented:

> [The history of the Sampling Theorem] reveals a process which is often apparent in theoretical problems in technology or physics: first the practicians put forward a rule of thumb, then the theoreticians develop the general solution, and finally someone discovers that the mathematicians have long since solved the mathematical problem which it contains, but in "splendid isolation".

However, it could be that Wiener did not attach much importance to the sampling theorem because he was not in the mood to abandon the realm of analog signals, which was the most natural and comfortable setting for him. His works with Lee and later with Bigelow belonged to the analog world, and he only switched to the digital case when he saw it as completely necessary, for example, introducing a discretization process to work with the Wiener–Hopf equations that appear in his prediction theory, giving rise to what we currently know as the "Wiener filter."

A main application: Computation of analog frequencies

The sampling theorem allows transforming arbitrary analog signals into vectors. It is a big step, insofar as it facilitates the discretization of all types of signals, with the promise that in the process no information will be lost.

In both the analog and digital cases, there are two versions for each signal. The first one is the version in the time domain — or space domain, if the signal depends on more than one variable — and the second one is the representation of the signal in frequencies. In most physical and engineering applications, researchers show a particular interest in frequency behavior, which best reveals the studied signal's physical properties. However, computing analog frequencies is very difficult for several reasons. To begin with, it requires the calculation of certain integrals that are often difficult to evaluate. In addition, there is often no analytical expression of the signal in the time domain, which makes everything enormously complicated.

So, why not use the frequencies calculated by the discrete Fourier transform of the digital version of the signal? The sampling theorem could be applied to convert the study of the analog signal into a mere analysis of its vector of samples. Next, we would calculate the frequencies detected by the discrete Fourier transform and ask ourselves if they really describe the behavior in frequencies of the original analog signal. Are they uniform samples of the Fourier transform of the analog signal?

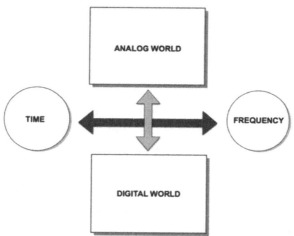

The truth is that the discrete Fourier transform of the vector that results from taking uniform samples of the analog signal $x(t)$ does not by itself adapt well to the world of analog frequencies. However, it will have a straightforward interpretation in terms of analog frequencies as long as during its computation we make some slight natural modifications.

Indeed, suppose that $\mathbf{x} = (x_0, x_1, \ldots, x_{N-1})$ with $x_k = x(kh)$ for a certain sampling period $h > 0$ and a certain analog signal $x(t)$. If the samples have been taken in another pattern, it will be necessary to rearrange the vector $\mathbf{x}$ to verify these equalities. Then, the discrete Fourier transform of $\mathbf{x}$ contains the information in analog frequencies of the original signal, although its identification involves some additional work. First, we have to rearrange the frequencies.

The elements of $\mathbb{C}^N$ are interpreted as periodic signals of period N defined on $\mathbb{Z}$. Therefore, the same happens with the components of $\mathbf{y} = dft(\mathbf{x}) = (y_0, \ldots, y_{N-1})$. This means that in this context the high frequencies are in the center of the vector and the low frequencies on its extremes. Taking this into account, it is clear that to graphically represent the discrete Fourier transform, the data must be rearranged by exchanging the first half of the vector with the second, sliding the first half to the right and the second to the left. In this way, the center of the new vector will correspond to the frequency 0, while the negative frequencies will be on its left and the positive ones on its right.

Moreover, the graphical representation of the data requires certain care. For example, the value x_k is not placed on the abscissa of value k but on the abscissa whose value is kh. Also, for the discrete Fourier transform, we must calculate the value of the fundamental frequency f_1. Now, if the sample period is h, the sample rate used will be $F_s = 1/h$. By the sampling theorem, this must be at least twice the maximum frequency involved in the signal. Therefore, with this sampling frequency, we will not be able to obtain information about the possible frequencies involved in the signal for frequencies higher than $1/(2h)$. Also, if the original analog signal contains frequencies above $1/(2h)$, the data provided by its digital version will only be useful for recovering an alias of the original signal. In fact, the discrete Fourier transform only gives us information about the frequencies between $-1/(2h)$ and $1/(2h)$. And as we only have N values that represent equispaced samples of the analog Fourier transform, it turns out that we must place the fundamental frequency at $f_1 = 1/(Nh)$ and the values of the discrete Fourier transform must be represented between the points $-(N/2)f_1 = -1/(2h)$ and $(N/2)f_1 = 1/(2h)$. Furthermore, the values y_k of the discrete Fourier transform are scaled by multiplying by the sampling period h. The following two figures show the effect of all these reorganizations and the scaling of the DFT for a concrete example.

The changes just explained provide a value of the discrete Fourier transform that is generally quite close to the true value of the analog Fourier transform of the original signal. There are however certain limits that must be taken into account.

We have already observed that the sample rate limits the bandwidth of the signals to be considered. If the signal is not band-limited

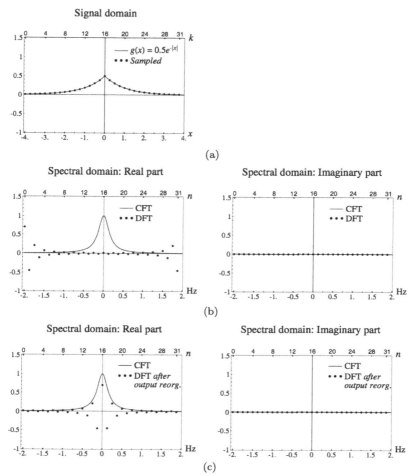

(a)

(b)

(c)

Example of sampled analog signal (a) and the DFT of the samples, first with no reorganization of the outputs (b) and then with the outputs reorganized (c). The CFT is the continuous Fourier transform of the original analog signal.

or the sampling frequency is smaller than the Nyquist frequency, we will not be working with the original signal but with an alias. Furthermore, even though the original signal is band-limited, and even though we have sampled with the Nyquist frequency, we will have problems because the sampling formula requires an infinite set of uniformly spaced samples, and our vector **x** is always finite. Therefore, in general, there is an inevitable mismatch in the results, called

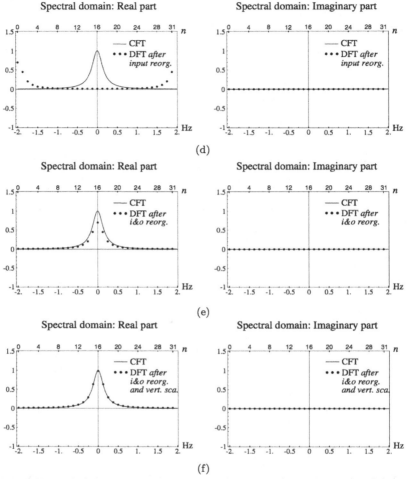

Continuation of the example above. Now, inputs are reorganized in (d), then both inputs and outputs are reorganized (e) and, finally, in (f) the outputs are also scaled. Only in this last case, the computed DFT can be considered a good approximation to a vector of samples of the Fourier transform of the original (analog) signal.

"spectral leakage," which is impossible to avoid. In any case, if we want to be sure that the discrete Fourier transform of the sample vector is near the frequency behavior of the original signal, oversampling is with few exceptions essential.

Shannon's Theory of Communication

Shannon's initial goal when he wrote his 1948 article was to understand and model, from a mathematical point of view, the different forms of communication that exist. When we communicate something, we transmit certain information from one point to another. Thus, when talking on the phone, we transmit speech from a sending terminal to a receiving terminal. The telegraph, for its part, allows the transport of text messages thanks to Morse code, and television does the same with sound and moving images. On the other hand, nature has its own mechanisms for transmitting information. For example, cells can build proteins, thanks to DNA, which transmits genetic information. And bees communicate to their companions the route to follow through a kind of dance whose code must be deciphered.

The ways of transmitting information are so diverse that it is extraordinarily surprising that someone found a mathematical model common to all of them, which also allowed the quantitative study and control of communications. That person was Shannon. His genius, which made his contribution so valuable, lies in the fact that his model was not intended to be restricted only to the forms of communication known until the time of publication but instead had a general abstract character that could be applied to the communication mechanisms of the future. It was a valid model for what was still unknown! Indeed, Shannon's ideas are used today in subjects that were unimaginable at the time, such as mobile telephony.

The scheme proposed by Shannon to summarize all communication processes was simple but powerful:

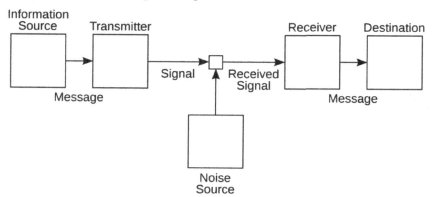

This simple diagram reduces to its essence — by way of abstraction — every act of communication. To begin with, we start from a source of information, which produces a message to be transmitted. To achieve this, there must be a transmitter, which takes the message and, through a specific mechanism, transforms it into a signal, which will be sent to the receiver. From the outset, the objective of the transmitter is to modify the original message without losing the information it carries so that it can move through a communication channel or medium. For example, in the case of telephone communication, the original message is speech, which must be transformed in some way into an electrical signal capable of traveling through a cable, the communication channel. The process of converting speech into the corresponding electrical signal is called "encoding" and is carried out by the telephone terminal from which it is spoken. The ultimate goal of encoding is to fit the original message to the channel. Once inside this, the message can suffer various distortions or disturbances caused by the physical process of transmitting the signal. This is the noise introduced into the communication system, which if excessive could prevent proper communication between the sender and the receiver. At the end of the process, the receiving device collects the received signal — the result of the sum of the transmitted signal and the noise introduced into the system — and decodes it, reversing the action carried out by the transmitter. In this way, it transforms the signal into the message for the recipient. The ultimate goal is to make the received message coincide with the one sent, trying to make the process as accurate as possible.

Shannon's model does not go into detail about the type of signal to be transmitted, the nature of the communication channel used, or the meaning of the message transmitted. Precisely by proceeding in this way, Shannon managed to offer a model general enough to encompass the different communication systems created through history, including those yet to be discovered. In addition, by decomposing the communication process into key elements, he made possible the independent study of each one of them.

The paper appeared in 1948 in the *Bell Systems Technical Journal*, a research journal accessed by communications engineers — electrical engineers at the time — and also read by some physicists and mathematicians. However, its readers were quickly aware that the topics discussed were of general interest. One of them,

Warren Weaver (1894–1978), convinced Shannon of the need to display the results in a format suitable for a wider audience. One year later, they published a joint book entitled *The Mathematical Theory of Communication*. The apparently superficial change in the title indicates that the text was not simply "a" proposal but, on the contrary, represented "the" proper way to deal with communication problems. That same year, another of Shannon's papers also appeared, *Communication in the Presence of Noise*. He focused his attention on recovering a signal (or a message) when it is disturbed by the addition of Gaussian noise while passing through the transmission medium. This led him to formulate and demonstrate the Shannon–Hartley formula, which relates the capacity of a channel to the bandwidth allowed by it and the signal-to-noise ratio, which results from dividing the power of the transmitted signal by the power of the noise that affects the channel (all these concepts will be explained later on; for now, suffice it to say that this is a result of enormous value for communication theory):

$$C = W \log_2 \left(1 + \frac{S}{N}\right).$$

Indeed, in the aforementioned works, in addition to proposing his model for the transmission of information, Shannon demonstrated several results that define the precise limits within which any communication system will work. These results, known today as "Shannon's theorems," marked the path to follow from then on. Moreover, they showed that there are certain limits for the transmission of signals of any kind through a communication channel, which depend on its nature and the source of information from which it is fed. Finally, they also showed that these levels are always achievable. However, his work did not give explicit solutions to the problem. On the contrary, the search for it would become the fundamental engine for the advancement of both the newly created theory and engineering practice, and in particular the design of different technologies related to the transmission of information.

In what follows, we provide a brief introduction to communication theory as introduced by Shannon. We will also explain the concept of Faster-than-Nyquist signaling, which has acquired great relevance in the second decade of this 21st century. Finally, we will briefly explain the controversy that existed between Shannon and Wiener regarding the paternity of information theory.

Information Sources

We already know that signals with a physical origin can be discrete-time registered without losing information. In particular, for one-dimensional signals, it is possible to obtain a good representation by recording a vector of samples $\mathbf{x} = (x_0, \ldots, x_N)$, where $x_k = x(kh)$ is, for each $k \geq 0$, a real number. Given that the physical devices in which this information is stored have to work with a level of precision determined in advance — as is the case with the devices used to take measurements of the signal — the values are usually also quantized. Indeed, it is impossible, for example, to store all the digits of numbers like $\pi, e, \sqrt{2}, \sin 1, \tan(2), \ldots$. For this reason, a specific format is usually set for the numbers with which a computer works. In this way, the set of numbers to be used is drastically reduced since it is no longer infinite. In reality, and although at first glance it might seem otherwise, this is not an insurmountable restriction. All measurable physical quantities are bounded by numbers that we can compute. And the number of significant digits required in the vast majority of applications — even those in very specific subjects, such as quantum mechanics or electroencephalography, where the quantities to be worked with are incredibly tiny — is also limited. Any computer can handle, with the necessary precision, all the valuable calculations both for our daily life and for scientific research, regardless of the branch of knowledge involved. Therefore, it is natural to assume that a finite set of discrete symbols expresses each value. In particular, we can always write the values of each sample x_k in base 2 so that the complete signal will be determined by a (very long) sequence of zeros and ones. This is important because, as we will see, once digitized, the information can be preserved and transmitted without introducing errors, which is excellent news for anyone who wants to configure a communication system.

Discrete sources of information

There are essentially three types of sources: discrete sources, analog sources, and discrete-time sources with analog values. The output of a discrete source is a sequence of symbols from a known discrete alphabet $\mathcal{A} = \{S_1, \ldots, S_h\}$. The output of an analog source is a function (a waveform) whose values are not discretized. The output of a discrete-time source with analog values is a sequence of real numbers. In this paragraph, we pay attention to discrete sources and their entropies.

Communication theory completely leaves out the meaning of the information carried by a communication channel. It only deals with the basic symbols used to transport it and their treatment throughout the communication process. The meaning of the symbols is set separately by the user. This was expressed with total clarity by Robert Fano (1917–2016), one of the first promoters of Shannon's theories, in an interview in 2001:

> [Information theory] is a theory about the transmission of information, not about the information.

To this claim, we must add that information theory is also concerned with precisely defining the *amount* of information contained in the messages generated by an information source. Instead of being related to the meaning of the messages, the mathematical concept of information is linked to the amount of surprise they contain, which allows information to be associated with the concept of a bit. To address it, we need to complete the definition of "source of information" by adding a little probability. This is done in a very simple way: It is enough to assume that each source of information produces its messages by randomly choosing the symbols that compose them. According to this extended concept, a discrete information source is a random process in which the symbols to be transmitted from a certain list, known as "alphabet," are chosen at successive instants of time.

We therefore approach the sources of information in probabilistic terms. It is all about assuming that, before receiving a message produced by a specific source of information, we have certain knowledge about the type of message that the source can emit, but we do not know exactly what the result will be. For example, if we know that the source emits texts in Spanish, it is unlikely — though not impossible — that the message contains words in another language. Shannon thought that one way to measure the informational content of a message in this context would be to calculate somehow the surprise it causes us. Within this logic, the information contained in a specific message is associated with the degree of predictability of the specific message we will receive. If before receiving it, we know with complete certainty what the message will be, it will not be informative at all. Receiving it will add nothing to what we already know. If, on the other hand, it is very unlikely that we can predict the message, receiving it will provide a lot of information. To formulate this property in mathematical terms, Shannon proposed that, in the context of a random experiment, the information provided by an event is given by the formula

$$I(A) = \log_2 \frac{1}{P(A)} = -\log_2 P(A) \text{ bits,}$$

where the bit is the chosen information unit; it represents the information obtained when making a choice between two events of probability $1/2$, since $P(A) = 1/2$ implies that $\log_2 \frac{1}{P(A)} = \log_2 2 = 1$.

For example, suppose that our source of information is a coin that is tossed so that the symbol that it emits each time is a head or a tail. If the coin is not rigged, each time we toss it, the source chooses both symbols with equal probability, and the result provides us with one bit of information each time. However, if the coin is rigged and the probability of heads is 0.9, while the probability of tails is 0.1, then the "tails" result is much more informative than the "heads" result. Finally, if the coin has two heads (or two tails), the experiment of tossing it is not informative at all because we know what the result will be before tossing the coin.

This way of defining the information is consistent with the idea that the information should be additive. If the source first emits a message M_1 and then independently emits another message M_2, the information provided by the successive appearance of both messages — which we can model as the double message $M_1 M_2$ — should be the sum of the information provided by both messages. And, indeed, this is what happens since if the messages are emitted independently, the probability of the new message is given by the expression $P(M_1 M_2) = P(M_1)P(M_2)$, and therefore

$$\begin{aligned} I(M_1 M_2) &= -\log_2 P(M_1 M_2) \\ &= -\log_2 P(M_1) - \log_2 P(M2) \\ &= I(M_1) + I(M_2). \end{aligned}$$

When describing sources of information as random processes, the appropriate mathematical objects for their formalization are random variables. For example, if the alphabet associated with our source of information is given by the finite set of symbols $\mathcal{A} = \{S_1, \ldots, S_N\}$, then we can associate with the source each time it emits a symbol a random variable X that takes values in the set of natural numbers $\{1, \ldots, N\}$, with $X = i$ when the source emits the symbol S_i. In fact, from now on, we will write indiscriminately $X = i$ or $X = S_i$. Therefore, at each moment, the source emits a symbol, we can completely identify said emission with a random variable, while the properties of the source will be determined by those of the random variables that define it. Now, all the properties of a random variable that take values in the set $\{1, \ldots, N\}$ are determined by the values

$$p_i = P(X = i), \quad i = 1, \ldots, N,$$

which define its probability distribution.

Thus, the information source is described by a succession of random variables $X_1, X_2, \ldots, X_m, \ldots$ in which the value of X_i completely determines the symbol chosen by the source in the i-th position of the message to be transmitted. If all these variables act independently of each other and are uniformly distributed, we will say that the information source is memoryless. It is clear that these types of sources must be studied paying attention to the probability distribution shared by the random variables that define the source. The information provided by the emission of the symbol S_i will be given by the formula

$$I(S_i) = -\log_2 P(S_i) = -\log_2(P(X = i)) = -\log_2 p_i,$$

while the entropy of the source, which is the average information per symbol emitted by the source, will be

$$H(\mathcal{F}) = H(p_1, \ldots, p_N) = \sum_{i=1}^{N} p_i I(S_i) = -\sum_{i=1}^{N} p_i \log_2 p_i$$

and is measured in bits/symbol. For example, if the symbols are binary digits (binits), the entropy is measured in bits/binit. If the source of information chooses at random between the values $\{0, 1\}$ and the probability of 0 is p, then the value of the entropy of the source is the function $\varphi(p) = -p \log_2(p) - (1 - p) \log_2(1 - p)$, whose graph appears in the following figure.

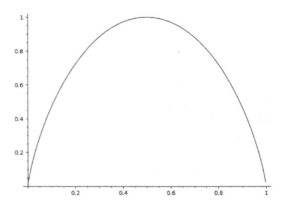

It is clear, then, that for memoryless binary sources — that is, memoryless sources whose alphabet consists of only two symbols — the average information per symbol will be at most 1, although the information provided by one of them (a binit) can exceed unity.

It should be noted that, by construction, we can substitute the memoryless source with the random variable that defines it. Thus, we can indiscriminately speak about the "entropy of the source," the "entropy of the random variable" that defines the source or, in general, the "entropy of a random experiment" that has a finite number of possible outcomes. We in turn identify all these measures with the function $H(p_1, \ldots, p_N) = -\sum_{i=1}^{N} p_i \log_2 p_i$, which is applied to any vector $(p_1, \ldots, p_N)$ of non-negative numbers that defines a probability distribution — that is, whose sum is equal to 1.

The sources of information considered so far are usually too simple. It is not reasonable to assume that the symbols emitted by a source are statistically independent of each other. For example, it is well known that every natural language has a good dose of redundancies that cause, in the case of written language, the appearance of the different pairs of letters to follow a certain pattern of probabilities. Thus, in Spanish, the letter "u" will appear after the letter "q" with high probability when compared to the probability of any other letter. And this is just a tiny part of the statistical information that we know about Spanish. If we want to establish a source model that generates written Spanish, we should take such facts into account, and also punctuation marks, spaces, parentheses, and hyphens, among others. It is evident that the best option is no longer a memoryless source.

In this context, it is convenient to use the so-called "Markov sources" conceived by A. A. Márkov (1856–1922). This Russian mathematician proposed to consider sequences of random variables in which the value of each variable may depend on the results of some of those that precede it. Therefore, we will say that the source described by the sequence of random variables $\{X_n\}_{n=0}^{\infty}$ is a Markov source of order m if the result of the variable X_n depends on the values of the m variables that precede it, $X_{n-1}, X_{n-2}, \ldots, X_{n-m}$, but is independent of the rest of the variables of the sequence. In the simplest case $(m = 0)$, we have the memoryless sources and we are not talking about the Markov property. The first important case is $m = 1$, which we normally call a "Markov source." The mathematical way to state that $\{X_n\}_{n=0}^{\infty}$ is a Markov source of order m is to impose the equations

$$P(X_n = y | X_1 = x_1, X_2 = x_2, \ldots, X_{n-2} = x_{n-2}, X_{n-1} = x_{n-1})$$
$$= P(X_n = y | X_{n-m} = x_{n-m}, \ldots, X_{n-1} = x_{n-1})$$

Furthermore, it is usual to impose the additional condition of homogeneity, which consists of assuming that the previous probabilities do not depend on n. Thus, if the alphabet used is $\mathcal{A} = \{S_1, \ldots, S_N\}$, then the Markov source of order 1 is determined by the joint knowledge of the probability distribution of the initial variable X_0 and the matrix of transition probabilities $P = (p_{i,j})$, where

$$p_{i,j} = P(X_n = S_j | X_{n-1} = S_i).$$

This is so because if $\mu^{(n)} = (p_1^{(n)}, \ldots, p_N^{(n)})$ is the vector that defines the probability distribution of the variable X_n — that is, $p_i^{(n)} = P(X_n = S_i)$ for $i = 1, \ldots, N$ — then $\mu^{(n)} = \mu^{(0)} P^n$. Of course, the more we increase the order m of the Markov source, the more complex the calculations become, but more flexible models are also achieved, with greater capacity to emulate the patterns that we observe in natural language.

Shannon came to use these ideas and, after analyzing a random number table with 100,000 entries — very similar to the tables of logarithms that were used before the advent of calculators and with which every high school student was familiar — he generated various samples of written text that, as the order of the Markov random source increased, became more and more similar to English. These techniques were later applied in natural language processing and are now one of the first automatic language generation models explained in computational linguistics courses. In addition, we can find them implemented in numerous applications, such as Google or WhatsApp, among others. With certain variations, they are also used to facilitate communication for people with disabilities.

We only have explained the concepts related to information sources for memoryless sources. Still, all these results and concepts also apply to Markov sources of arbitrary order, and many of them also extend to other types, such as stationary and ergodic sources.

Axiomatic Definition of Entropy

The formula that defines entropy is not an arbitrary choice. In his 1948 paper, Shannon motivated his definition on the basis of several simple properties. In 1957, the Russian mathematician A. I. Khinchin (1894–1959) showed that Shannon's entropy is characterized — except for change of scales — by the following four properties:

(i) The function $H(p_1, \ldots, p_N)$ reaches its maximum in the uniform distribution (that is, when $p_1 = \cdots = p_N = \frac{1}{N}$), so we always have that

$$H(p_1, \ldots, p_N) \leq H(\frac{1}{N}, \ldots, \frac{1}{N}).$$

(ii) If X, Y are independent discrete random variables and we consider the variable (X, Y), which results from doing first X and then Y, then

$$H(X, Y) = H(X) + H(Y).$$

(iii) $H(p_1, \ldots, p_N) = H(p_1, \ldots, p_N, 0).$
(iv) $H(p_1, \ldots, p_N)$ is a continuous function.

The first property indicates that there is always the highest level of uncertainty for the experiment in which none of the elementary events has an advantage over the rest. The second ensures that the information is additive for independent events. The third tells us that adding impossible events to an experiment never increases — or decreases — the information we have about it. Finally, continuity is a minimum requirement for whatever magnitude we want to define.

The axiomatic characterization of entropy is a typical example of the use of functional equations techniques, i.e., equations that do not involve certain processes of classical analysis, such as the use of integrals or derivatives. After the work of Khinchin, the axiomatic characterization of the different measures of information or entropy was studied in great depth, especially by the Hungarian mathematicians J. Aczél and Z. Daróczy, who published in 1975 an important monograph dedicated exclusively to characterizing the different ways in which it is reasonable to measure information.

Knowing the entropy of information sources is fundamental to the objectives of communication theory. That the source $\mathcal{F}$ has entropy H means that on average each symbol emitted by the source contains H bits of information. Therefore, if it emits r symbols per second, the channel to which it is connected receives rH bits per second. For everything to work correctly, we will need that this amount does not exceed the capacity of the channel, which measures the amount of information in bits per second that a channel is capable of admitting and processing. Therefore, entropy is an important information measure, where we take into account the global behavior of the stochastic process represented by the information source.

Transmitters

We have already introduced discrete information sources and have shown some of their properties. The next prominent element in Shannon's scheme is the transmitter, which modulates and encodes the information before it is fed into the communication channel.

There are three compelling roles for this process. The first is to try to compress the information provided by the source through a process called "source encoding." The second is to adapt the source's output so that the channel can receive and transmit it correctly, and this is achieved with the modulation of the signal emitted by the source. The third, no less important, is to minimize possible transmission errors.

Signals modulation

The first step to enable the transmission of the messages produced by a source, whatever its nature, is to adapt the source's output to the communication channel. This pursues the elementary objective that the channel accepts as input the symbols emitted by the source while it seeks to avoid overlaps and interferences between different emitting sources, if the channel can accept several of them simultaneously.

For example, microphones allow us to transform a sound signal into an electrical one that can then be transmitted over a cable when we speak on the phone. In radio waves, each station receives the same type of sound signal before sending it out for listeners to listen to its channel. Therefore, the signals they send must share the same frequencies, and they must be in the same area of the electromagnetic spectrum. If these signals were not modulated in some way to travel on separate frequency bands, radio sets would be unable to filter the different channels and would receive all the overlapping waves in the same area of the spectrum, which would cause interference with each other. For this reason, the message to be broadcast is modulated using a carrier signal that shifts the signal to the area of the electromagnetic spectrum assigned to the channel. As the different assigned zones do not overlap, the receiver device must possess a filter that passes the frequency band corresponding to the emission of the chosen channel. The same ideas apply to all communication systems that take advantage of electromagnetic waves, including television, mobile telephony, and satellite networks, among others.

Suppose for a moment that the signal we want to modulate is the function $x(t)$ which has bandwidth $W < \infty$. Normally, we will take a much higher frequency f_0 and seek to build a new signal $s(t)$ — the modulated signal — whose frequencies are concentrated around f_0 (and $-f_0$), with the important property that it must be possible to recover the signal $x(t)$. There are several ways to build the signal. The simplest one is by modulating the amplitude of a carrier signal. Specifically, we can define

$$s(t) = A \cdot (1 + m_{AM} x(t)) \cos(2\pi f_0 t),$$

where A, m_{AM} are positive constants and we have forced that $1 + m_{AM} x(t) > 0$ for every $t \in \mathbb{R}$. The second method, more complex because it is nonlinear, consists of modulating the frequency instead of the amplitude of the carrier signal. To do this, you take

$$s(t) = A \cos \left(2\pi f_0 t + m_{FM} \int_{t_0}^{t} x(\tau) d\tau \right).$$

From a theoretical point of view, amplitude modulation is simpler, mainly because it is a linear process and it has a more obvious justification. However, it can be shown that in practice frequency modulation produces better results because it increases the signal-to-noise ratio.

If the information source is discrete, it is also possible to use classical Fourier analysis techniques to modulate it, which in fact is often necessary. For example, when we send information over the Internet, our computer — the source of information — produces a binary source, which, however, can circulate on a cable or as electromagnetic waves, channels that are naturally analog.

The simplest example is that of a binary source whose alphabet is $\mathcal{A} = \{-1, 1\}$. The source outputs are sequences (a_n) in which $a_n \in \{-1, 1\}$ for all n. It is not difficult, therefore, to construct from these values an analog signal that allows us to fully identify the sequence that has been used. For this, it is enough consider a signal

$$s(t) = \sqrt{E_s} \sum_{n} a_n h(t - nT),$$

where $h(t)$ is what is called a "pulse." Usually, it is a function that satisfies the orthogonality condition

$$\int_{-\infty}^{\infty} h(t - kT) h(t - mT) dt = 0, \text{ if } m, k \in \mathbb{Z} \text{ and } m \neq k,$$

satisfies $h(0) = 1$, has energy $\int_{-\infty}^{\infty} |h(t)|^2 dt = 1$, and has a Fourier transform that concentrates almost all its energy in a symmetric interval with respect to 0, being negligible outside of it.

For example, in the figure below the graphs of three frequently used pulses have been drawn: the rectangular pulse

$$h(t) = \frac{1}{\sqrt{T}} \text{rect}(t/T),$$

the triangular pulse

$$h(t) = \begin{cases} \sqrt{3/T}(1 - 2|t|/T) & |t| \leq T/2 \\ 0 & |t| > T/2 \end{cases},$$

and the raised cosine

$$h(t) = \begin{cases} \sqrt{2/3T}(1 + \cos(2\pi t/T)) & |t| \leq T/2 \\ 0 & |t| > T/2 \end{cases}.$$

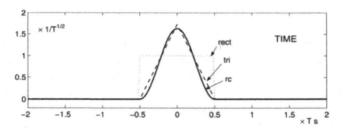

The following figure shows the corresponding Fourier transforms:

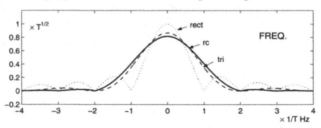

The figure shows how the smoothness of the pulse affects the decay of the Fourier transform: the smoother, the faster the decay and, therefore, the better treatment the pulse train has in the frequency domain. Finally, the figure below shows how the technique of using a pulse train allows a digital signal to be transformed into analog for transmission through an appropriate channel:

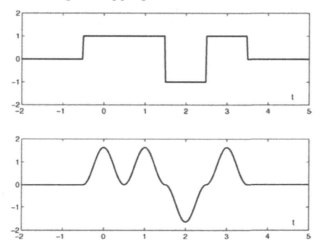

In this case, two trains of pulses have been considered, both carrying the sequence $\{1, 1, -1, 1\}$; one is made up of rectangular pulses, while the other is based on the raised cosine.

If we want to shift this — or any other — pulse train in frequency, it is enough to consider a signal of the type

$$s(t) = \sqrt{2E_s} \left(\sum_n a_n h(t - nT) \right) \cos(2\pi f_0 t).$$

Source and channel encoding

Let us suppose that we have a binary source whose alphabet is $\mathcal{A} = \{0, 1\}$ and that the communication channel has no problem accepting as input the elements of $\mathcal{A}$. Suppose also that the channel uses the same symbols, so its output is zeros and ones. Now, there is a probability $p \leq 1/2$ that an output is wrong and consequently the channel shows a 0 when a 1 enters or shows a 1 when a 0 enters (if $p > 1/2$, changing all the outputs we would achieve that the new channel satisfies the restriction $p \leq 1/2$). Will it be possible to reduce the probability that the channel produces a wrong output? An easy way to achieve this is simply to introduce some redundancy into the system. For example, when the source emits a symbol, instead of introducing it only once in the channel, we could send the same symbol repeated three times and then look at the output and choose the symbol that has appeared with the greatest frequency. Then, the probability of being wrong with this method will be equal to

$$\binom{3}{2} p^2(1-p) + \binom{3}{3} p^3,$$

which is always smaller than p. For example, when $p = 1/3$, the previous expression is equal to

$$3\frac{1}{9}\frac{2}{3} + \frac{1}{27} = \frac{7}{27} < \frac{1}{3}.$$

Moreover, it is not difficult to show that by forcing the number $2n+1$ of times the symbols are repeated to be large enough and always choosing the symbol that is repeated the most times, we can reduce the probability of error as much as we want, ensuring a completely reliable transmission.

This, transforming the outputs of a given source into blocks of symbols that may be a different source, is called "encoding," or just coding. In fact, it is a special type of coding, named word-block coding.

We have just shown that it is always possible to achieve a coding that reduces transmission errors as much as desired, although we have done so is a rough way, which almost paralyzes the transmission of information and is useless in practice. There is, however, something correct in this example: It points in the right direction, since it tells us that the price to pay if we want to guarantee reliability is slowing it down. One might wonder, then, by how much?

The answer is that to achieve reliable transmission it is not necessary to reduce the transmission rate to zero, but rather it must be reduced to a number strictly positive, which is the capacity of the transmission channel. To achieve this, it is necessary to carry out an efficient coding, which allows the data produced by the source to be transmitted reliably without much sacrificing the transmission speed. This of course is closely related to the problem of efficient representation — in other words, compression — of information: If we want to maintain the speed of transmission and in addition we want to introduce the redundancies that allow a reliable decoding, it is necessary first to compress the information to its essential content — a process that we will call "source coding." Then, if necessary, the output of the new, already compressed source will have to be recoded to adapt it to the capacity of the channel — this is known as "channel coding."

One way to address these issues is to use block encoding, which works as follows: The source is allowed to output a block of symbols of a predetermined size, and each block thus obtained is assigned a sequence of symbols, known as the "codeword," which is prepared in such a form that the channel can transmit it. This association of symbol blocks with code words must be such that when the information is received on the other side of the channel, it can be decoded quickly and reliably.

Block coding and data compression

Let's see how block encoding of an information source is approached and how it can be applied to data compression. We start from two sets of symbols: the source alphabet $S = \{S_1, \ldots, S_N\}$, used by the source, and the code alphabet $\mathcal{A} = \{a_1, \ldots, a_D\}$, which is used by the channel. The initial objective is to associate each message emitted by the source with a message that the channel can accept, a process that must be carried out by optimizing certain values. Specifically, it is intended that the average length of the encoded messages is minimal, which will facilitate the flow of information from the source to the channel. To produce one of these codes, a code word is associated with each letter of the font: $\phi(S_i) = X_i = a_{i,1} a_{i,2}, \ldots, a_{i,\ell_i}$, which will have length $\ell_i \geq 1$, where $a_{i,j} \in \mathcal{A}$ for all i, j. There are ways to guarantee that this association can be decoded uniquely, so that when a message $m_1 m_2 \cdots m_h$ made up of the letters of the code alphabet is observed, there is an algorithm that tells us if the message is of the form $X_{i_1} X_{i_2} \cdots X_{i_k} = \phi(S_{i_1}) \phi(S_{i_2}) \cdots \phi(S_{i_k})$ or, on the contrary, the given sequence of symbols cannot be written in this way. Some simple examples of encodings are Morse, Baudot, and ASCII codes.

Shannon showed that if we apply one of these codes using a code alphabet of r letters for memoryless sources of information, then there exists an encoding such that

$$\frac{H(S)}{\log_2 r} \leq \bar{\ell},$$

where $H(S) = -\sum_{i=1}^{N} p_i \log_2 p_i$ is the entropy of the source and $\bar{\ell} = \sum_{i=1}^{N} p_i \ell_i$ is the average length — that is, the expected value of the length — of the code words used. Moreover, there are uniquely decodable codes for which the inequality

$$\bar{\ell} < \frac{H(S)}{\log_2 r} + 1$$

is also verified. So, when we take one of these codes, the inequalities can be rewritten as

$$H_r(S) \leq \bar{\ell} < H_r(S) + 1,$$

where

$$H_r(S) = \frac{H(S)}{\log_2 r} = -\sum_{i=1}^{N} p_i \frac{\log_2 p_i}{\log_2 r} = -\sum_{i=1}^{N} p_i \log_r p_i$$

denotes the entropy of the source, measured with r-ary units rather than binary units (bits). In particular, an encoding exists such that the average length of the words of a r-ary block code never falls below the entropy $H_r(S)$. This result is known as "Shannon's First Theorem" or "Shannon's Source Theorem."

Note that the inequality $\frac{H(S)}{\log_2 r} \leq \bar{\ell}$ is quite natural. In the alphabet used by the channel — the code alphabet — there are r symbols to choose from, so each letter will carry at most $\log_2 r$ bits of information. Therefore, if we consider a code based on this alphabet, then the average of the information that each code word will provide will be $\bar{\ell} \log_2 r$ bits and each of these words will be linked to a single symbol of the source. As the code is uniquely decodable, this quantity cannot be less than $H(S)$, which represents the average number of bits per symbol of the source.

If instead of applying the code to the letters of the alphabet, we do it to blocks of letters, which is possible by simply considering the n-th order extension of the source $S^n = \{S_{i_1} S_{i_2} \cdots S_{i_n} : S_{i_j} \in S$ for all $i_j\}$, and applying the block code to that extension, the inequalities we obtain are

$$nH_r(S) = H_r(S^n) \leq \overline{\ell_n} < H_r(S^n) + 1 = nH_r(S) + 1$$

And, therefore,

$$H_r(S) \leq \frac{\overline{\ell_n}}{n} < H_r(S) + \frac{1}{n}.$$

This is important because it ensures that, using an r-letter code alphabet, we can get as close as we want to the value $H_r(S)$ by directly encoding blocks of words from the source. If $N \leq r$, then $H_r(S) \leq \log_r N \leq 1$, and therefore $R(S) = 1 - H_r(S)$ is a measure of the level of compression that can be achieved by encoding the source with an r-ary code.

For example, in the case that the source is binary and $r = 2$, Shannon's theorem guarantees that a binary file of size K generated by a source with entropy H can be compressed to a file of size at most HK, possibly with a large K, which implies a reduction of $(1 - H)K = RK$ bits. Since the entropy of a binary source is always ≤ 1, the statement makes sense: We are talking about a true compression. In addition, as we have already observed, the same idea can also be applied with non-binary sources. Let's take an example from digital photography. It is known that, after quantization in the JPEG format, the obtained files are composed of bytes — one byte is equivalent to 8 bits, which allows us to specify $2^8 = 256$ values, which are the gray levels of the pixels in the photo. We can assume that the byte with value 0 appears with probability 0.5 and bytes with values between 1 and 20 appear with probability 0.4. Let us assume — in the absence of more data — equidistribution of the values. That is, assume that each byte between 1 and 20 has probability $0.4/20$ and that the rest of possible values — which are the 235 numbers greater than 20 — appear with probability $0.1/235$. Due to the uniformity assumptions for the probability distribution we have just defined, the entropy of the modeled source will be greater than or equal to the entropy of the "real" source that generates the picture. In particular, the value of the entropy of the modeled source is

$$H = -0.5 \log_2 0.5 - \frac{0.4}{20} \log_2 \frac{0.4}{20} - \frac{0.1}{235} \log_2 \frac{0.1}{235} = 3.877\ldots$$

That is, each byte of the picture contains at most 3.87 bits of information. Typically, a 100-kB file could be printed with a suitable encoding using $100 \cdot 1024 \cdot 3.877 \approx 397000$ bits, approximately 48.5 kB.

Huffman codes

We owe D. A. Huffman (1925–1999) the construction of codes for which the average length of the code words is as short as possible. The way he got it was nicely reported in a paper published by the journalist Gary Stix in *Scientific American in 1999*.

In 1951 David A. Huffman and his classmates in an electrical engineering graduate course on information theory were given the choice of a term paper or a final exam. For the term paper, Huffman's professor, Robert M. Fano, had assigned what at first appeared to be a simple problem. Students were asked to find the most efficient method of representing numbers, letters or other symbols using a binary code. Besides being a nimble intellectual exercise, finding such a code would enable information to be compressed for transmission over a computer network or for storage in a computer's memory.

Huffman worked on the problem for months, developing a number of approaches, but none that he could prove to be the most efficient. Finally, he despaired of ever reaching a solution and decided to start studying for the final. Just as he was throwing his notes in the garbage, the solution came to him. "It was the most singular moment of my life," Huffman says. "There was the absolute lightning of sudden realization."

Up to that point, everyone had tried to assign code words to the alphabet of the source starting with the most frequent symbol — with which is associated the shortest code word — and moving on to the less frequents symbols, increasing the length of the code words, if necessary. And Huffman had been proceeding that way for months. However, he realized that the optimal code word could be achieved by moving in reverse direction.

Suppose we want to build one of these codes using a binary code alphabet $\mathcal{A} = \{0, 1\}$. The problem is trivial for binary sources. For a three-element font alphabet, with probabilities p_1, p_2, p_3, the assignment is done with a tree of the following type:

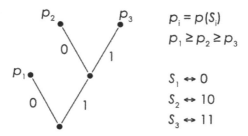

$$p_i = p(S_i)$$
$$p_1 \geq p_2 \geq p_3$$

$$S_1 \leftrightarrow 0$$
$$S_2 \leftrightarrow 10$$
$$S_3 \leftrightarrow 11$$

Note that we have placed the leaves — which do not have children — in the upper part of the tree, while the root is placed at the bottom, supporting the entire graph. The idea is to place the different probabilities on the leaves of the tree. The way to assign a code is to label the edges. Each node which is not a leaf has two edges; one (to which a 1 is assigned) that departs to the right and the other (to which a 0 is assigned) to the left. To decide the code that reaches a given leaf — and that, therefore, is assigned to the symbol of the source that has the probability associated with that leaf — we look to the unique possible path of the tree that connects the root to the leaf (in this direction) and write, in the order we meet them, the labels of the edges. In the case of the tree with three leaves, we have to decide how to place the probabilities on the leaves. The solution is simple: It is enough to place the highest probability on the leaf closest to the root, while it does not matter how the others are placed.

Suppose the problem for alphabets of N elements has been solved. Let's see how to use the solution obtained to solve the case of a source with an alphabet of $N + 1$ elements $\mathcal{A} = \{S_1, \ldots, S_{N+1}\}$, which have probabilities $p_1 \geq p_2 \geq \cdots \geq p_{N+1}$ — here we assume $p_i = P(S_i)$ for all i. To do this, we consider the solution of the problem when the source has an alphabet $\mathcal{A}^* = \{S_1, \ldots, S_{N-1}, S_N^*\}$ — where S_N^* is a new symbol, different from the rest of the symbols S_i, $i = 1, \ldots, N + 1$, with probabilities $p_i = P(S_i)$, $i = 1, \ldots, N - 1$, and $p_N^* = p_N + p_{N+1} = P(S_N^*)$.

Starting from the tree that defines the solution of the problem for $\mathcal{A}^*$, the solution of the problem with $N+1$ elements is obtained in the simplest way: It is enough to create two leaves connected to the node whose associated probability is p_N^*, and assign the corresponding probabilities p_N, p_{N+1} to these leaves as desired.

For example, when we apply this algorithm to the source

$$S = \{S_1, S_2, S_3, S_4, S_5\}$$

with probabilities

$$p_1 = \frac{1}{3}, p_2 = \frac{1}{5}, p_3 = \frac{1}{5}, p_4 = \frac{1}{6}, p_5 = \frac{1}{10},$$

we get the following tree:

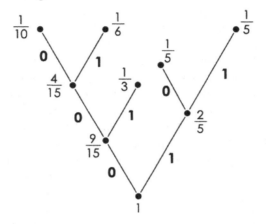

and therefore the associated Huffman code is

$$
\begin{aligned}
S_1 &\leftrightarrow 01 \\
S_2 &\leftrightarrow 10 \\
S_3 &\leftrightarrow 11 \quad . \\
S_4 &\leftrightarrow 001 \\
S_5 &\leftrightarrow 000
\end{aligned}
$$

In this case, the average length of the code words is $2.266\ldots$, while the source entropy is 2.22.

Huffman coding is not only applied in data compression — where it is used in algorithms such as JPEG, MP3, and PKZIP, among many others — but it is also used in cryptography, to build the so-called "one-way functions." These include the well-known SHA functions, which are applications that are easy to evaluate but very difficult to reverse and are used for electronic signatures and key creation.

We have explained how to use block codes to compress the information generated by a source. However, we have not yet defined in a formal way what we mean by "capacity" and we also have to explain how coding is used to ensure reliable transmission across a channel. Addressing these issues inevitably requires a good mathematical model of the information channel and, in particular, of noise.

Communication Channels

Suppose we have a discrete information source whose associated alphabet is $\mathcal{A} = \{S_1, \ldots, S_h\}$. The symbols shown here not only make it possible to compose messages aimed at a specific objective — such as, for example, expressing an idea in writing — but they also have to be introduced into a channel, which will carry them from one place to another or will preserve them in a physical storage device. It is worth noting that each time one of these symbols enters the communication channel, it takes a certain time to assimilate it, which may depend on the symbol. We call its value "symbol duration."

It follows that to model a channel that accepts as input the messages of the selected information source, it is necessary to know the durations $D = \{t_1, \ldots, t_h\}$ associated with the different elements of the alphabet $\mathcal{A}$. Thus, the information source emits its messages, which are ordered sequences of symbols $M = M_1 M_2 \cdots M_m$, where $m \geq 1$ is an integer and $M_i \in \{S_1, S_2, \ldots, S_h\}$ for all i. The duration of the message is the sum of the durations of the symbols that appear in the sequence that describes it.

Suppose we have a time interval $T > 0$ for the source to issue a message. In that case, we should calculate the number $N(T)$ of different messages that the channel can receive during that time, because the result provides us with a measure of the total information per second that the source can enter in the channel.

The simplest case where these calculations can be made is that of a memoryless binary source in which the duration — and the probability — of the two symbols of the alphabet is the same. That is, $\mathcal{A} = \{S_1, S_2\}$, chosen by the source with probability $1/2$ each $t > 0$. Since we have a time interval of size T, it contains $m = T/t$ symbols, and therefore the source can produce $N(T) = 2^m = 2^{\frac{T}{t}}$ different messages. If all are equally likely, the information transmitted by the source when it selects any one of them is given by $I = -\log_2 \frac{1}{2^m} = m$ bits, so it will be able to send $m/T = (\frac{T}{t})/T = \frac{1}{t}$ bits per second.

Note that this value matches $\frac{\log_2 N(T)}{T}$ for all $T > 0$. In general, this quantity will depend on T, and Shannon in his 1948 paper defined the capacity of a discrete noiseless communication channel as the limit

$$C = \lim_{T \to \infty} \frac{\log_2 N(T)}{T},$$

which represents the maximum information flow, measured in bits per unit of time, that the channel can assimilate.

When different durations are used for the different symbols of the alphabet, a precise computation of $N(T)$ may be difficult since to count the many ways in which a combination of alphabet symbols can fit exactly within a given time interval is not trivial. In any case, the problem can be reduced to the study of certain finite difference equations and, in many cases, C can be computed with high precision. For example, for a theoretical model of the telegraph, it can be shown that 12.936 bits/s is the maximum amount of information that a noiseless telegraph channel can process when receiving Morse code. Under these conditions, it is impossible to send messages at a speed that exceeds, for example, 13 bits per second, because the channel will collapse. Obviously, each channel will have its own specifications and, in a way, this is what we are talking about when we say that our internet service arrives with a certain bandwidth.

Discrete channels with noise

Once the source of information has produced a message, it must enter the communication channel to send it to the recipient. In the event that the channel must transmit discrete information, it will accept as input the symbols of the source alphabet $\mathcal{A}$, and it will transform them into the symbols of another — possibly different — alphabet $\mathcal{B}$, which will reach the receiver. If we are facing a perfect channel, without noise and without memory, it will not modify the messages that reach it or, in case the input and output alphabets are different, it will simply transform each input symbol into a unique one. However, all practical channels introduce noise or distort the signal delivered to them.

To deal with this problem, we introduce probabilities to model the noise as a stochastic process. Specifically, for each letter a_i of the input alphabet, we can distribute occurrence probabilities among the different letters of the output alphabet when we know that the input has been the chosen letter. That is, instead of assuming a deterministic behavior for the channel, we assume that each input sets a probability distribution among the outputs, which can be interpreted as a conditional probability. Thus, in addition to the input and output alphabets, our noisy channel model uses the conditional probabilities

$$q_{i,j} = P(b_j|a_i) := P(\text{the output is } b_j| \text{ the input was } a_i)$$
$$= \frac{P(a_i, b_j)}{P(b_j)},$$

where $\mathcal{A} = \{a_1, \ldots, a_N\}$ and $\mathcal{B} = \{b_1, \ldots, b_M\}$ are the input and output alphabets of the channel, respectively.

Note, however, that by forcing that the probability of each output depends exclusively on the input that the channel receives at the moment — and therefore does not depend on what has happened before — we are imposing that the channel lacks memory. As with the memoryless sources, this assumption is actually very restrictive. Although we will make our observations under this umbrella, the truth is that all the statements that we will make are also true in broader contexts.

Once it has been assumed that we are working with a channel without memory, it is clear that everything is described by the input and output alphabets and by the matrix $Q = (q_{i,j})$ that determines the different transmission probabilities between the letters of both alphabets.

We can, therefore, calculate several entropies. To begin with, we have the one from the source, $H(\mathcal{A}) = -\sum_{i=1}^{N} P(a_i) \log_2 P(a_i)$. On the other hand, taking into account that we know the values $q_{i,j} = P(b_j|a_i)$, we can compute the probability of the channel outputs by

$$P(b_j) = \sum_{i=1}^{N} P(a_i) q_{i,j},$$

and consequently we will have the entropy of the channel output

$$H(\mathcal{B}) = -\sum_{j=1}^{M} P(b_j) \log_2 P(b_j).$$

Furthermore, given that the output b_j has occurred, we can consider the possible influence of this fact on the probabilities of appearance of the input alphabet letters, which is done by calculating the conditional probabilities $P(a_i|b_j) = \frac{P(a_i,b_j)}{P(b_j)} = \frac{P(a_i)q_{i,j}}{P(b_j)}$. Associated with these probabilities, we define the conditional entropy

$$H(\mathcal{A}|b_j) = -\sum_{i=1}^{N} P(a_i|b_j) \log_2 P(a_i|b_j).$$

Once these entropies are known, a new concept is introduced, called "equivocation of $\mathcal{A}$ with respect to $\mathcal{B}$," whose formula is given by

$$H(\mathcal{A}|\mathcal{B}) = \sum_{j=1}^{m} H(\mathcal{A}|b_j) P(b_j).$$

Finally, we denote the mutual information between the input and the output of the channel by $I(\mathcal{A}, \mathcal{B})$. It is a measure that quantifies the extent to which the knowledge of the input reduces the uncertainty about the output and vice versa:

$$I(\mathcal{A}, \mathcal{B}) = H(\mathcal{A}) - H(\mathcal{A}|\mathcal{B}) = H(\mathcal{B}) - H(\mathcal{B}|\mathcal{A}) = I(\mathcal{B}, \mathcal{A}).$$

It is important to note that with our definition of the channel, the probabilities of appearance of the different output symbols depend exclusively on the probability distribution of the input source symbols, and on the matrix of probability transitions. Now, this matrix is what defines the response of the channel to the inputs. Therefore, it makes sense to find the probability distribution of the source that maximizes mutual information. The maximum value obtained in this way is what we know as channel capacity:

$$C = \max_{\{P(a_i)\}} I(\mathcal{A}, \mathcal{B}).$$

Binary Symmetric Channel

Among the binary communication channels, the binary symmetric channel is the simplest possible. We take $\mathcal{A} = \mathcal{B} = \{0, 1\}$ and assume that the probability of error in the transmission of each symbol is the same $p > 0$, so that the matrix probabilities of transmission of the symbols is

$$Q = \begin{pmatrix} 1-p & p \\ p & 1-p \end{pmatrix}.$$

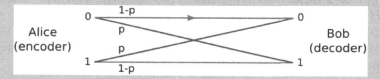

With these data, it is easy to calculate the capacity of the channel, which is $C = 1 - H(p)$, where

$$H(p) = p \log_2 \frac{1}{p} + (1-p) \log_2 \frac{1}{1-p}.$$

Let us now assume that the source emits r symbols per second and that the channel can transmit ρ symbols per second. This would mean that, on average, the source emits $rH(\mathcal{A})$ bits per second and the channel can support, also on average, ρC bits per second at most. Shannon's channel theorem guarantees that, if $rH(\mathcal{A}) < \rho C$, then for each $\varepsilon > 0$ there is a block encoding of the source and a decoding of the output such that, on average, the source symbols are represented by at most ρ/r characters. It also allows us to be sure that the probability of error in the transmission of each character — including the decoding process — is uniformly less than ε. Furthermore, if $rH(\mathcal{A}) > \rho C$, then there is a certain value $\varepsilon_0 > 0$ such that the average of the probabilities of error in the transmission of the different symbols of the source always exceeds ε_0, independently of the way the encoding is carried out.

Note that the parameter ρ used in formulating the theorem is the value that the channel capacity would have if the channel is noiseless and we know the durations of the source symbols. In other words, it is the capacity calculated in the previous section when $\mathcal{A} = \mathcal{B}$ and there is no noise or distortion. This is a value that we can calculate if we know the duration of the input alphabet symbols. In that context, we have that $I(\mathcal{A}, \mathcal{A}) = H(\mathcal{A})$, and the theorem states that we can send the information from the source reliably as long as the transmission speed is $r < \rho$, but that the channel will collapse if $r > \rho$. When introducing the idea of a noisy channel, this capacity is altered and must be modified by the mutual information of the input and output sources. Opposite to the channel without noise, we have a channel with noise in which the output b_j does not depend on the input, so that $q_{i,j} = P(b_j|a_i) = P(b_j)$ and, consequently, $P(a_i|b_j) = P(a_i)$ and $H(\mathcal{A}|b_j) = H(\mathcal{A})$ for all j. This leads to $H(\mathcal{A}|\mathcal{B}) = H(\mathcal{A})$ and $I(\mathcal{A}, \mathcal{B}) = 0$. It means that if the channel introduces so much noise that the outputs do not depend on the inputs, reliable transmission is impossible, which we could have expected.

Continuous channels: Shannon–Hartley formula

The case of continuous channels is different and requires a specific formulation. Although the sampling theorem indeed makes it possible to reduce, through time discretization and quantization of the signals, any analog signal to a digital transmission process, and therefore makes it possible to model that process with discrete information sources and discrete channels, it is also true that it would be desirable to have a formulation closer to the physics of the channels involved. In particular, we should be able to relate the bandwidth of the transmitted signal to the capacity of the transmission channel. After all, classical channels are physical devices — such as electrical or fiber optic cables — or electromagnetic waves. When we study their physical behavior, they have specific properties that we can take advantage of. For example, electrical transmission involves the appearance of a particular type of noise known as "white noise" or "Gaussian noise." Shannon was able to model the transmission through the physical channels that support noise of this type and prove, in this context, a theorem analogous to the channel theorem that we have exposed in the previous section. Let us explain how he did it.

We will focus on the transmission of one-dimensional analog signals. We are transmitting a function $x(t)$ that could represent, for example, a sound or an electrical potential. We assume that the signal is limited by a bandwidth $W < \infty$ and that we are interested in transmitting it during the time interval $[-T/2, T/2]$. So, this situation perfectly fits with the Slepian concept of band-limited, time-limited signals at level ε and, thanks to the sampling theorem, we can assume that the problem boils down to transmitting the samples $c_k = x\left(-\frac{T}{2} + \frac{kT}{n}\right)$, with $0 \le k \le n = N(W, T, \varepsilon, \epsilon)$, where k is an integer. Of course, every time we want to transmit a value c_k in a continuous channel even without noise, we must consider an infinite sequence. The same thing happens if we introduce noise — quantifying the signal, for example. Therefore, either of these two options gives rise to a source of infinite information, to which we cannot apply the assumptions of Shannon's channel theorem. The obvious conclusion is that it is necessary to impose the existence of noise and limit the power of the emitted signals.

Specifically, if we take a signal of finite bandwidth $\le W$ and time-limited to the interval $[-T/2, T/2]$ at level $\varepsilon > 0$, $x \in \mathcal{X}_\varepsilon$, and we set $n = N(W, T, \varepsilon, \epsilon)$, then $x(t)$ will be determined at level ε in the interval $[-T/2, T/2]$ by the vector $(c_1, \ldots, c_n)$, $c_k = x(-T/2 + kT/n)$. Moreover, the quantity

$$P = \frac{1}{n} \sum_{k=1}^{n} c_k^2$$

represents the average power of the signal in that interval. Therefore, the distance d from the point $(c_1, \ldots, c_n)$ to the origin of coordinates of the ordinary space $\mathbb{R}^n$ satisfies the relation

$$d^2 = nP.$$

In this way, if we limit the average power of the transmitted signals to a value $S < \infty$, the points $(c_1, \ldots, c_n)$ in space $\mathbb{R}^n$ associated with these signals must all lie within a sphere centered on the origin with radius $r = \sqrt{nS}$. For the same reason, if we allow the addition of a random noise of bounded power $N_0 < \infty$, the effect it will have on the received signals will be a displacement of the point $(c_1, \ldots, c_n)$

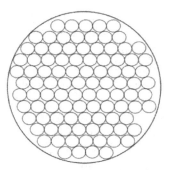

Sphere packing for the Gaussian channel.

to another in the sphere centered on it with radius $\sqrt{nN_0}$. It can be shown that the probability the new point is outside this sphere tends rapidly to 0 as n grows. In any case, all these points must live in the sphere whose center is the origin with radius $\sqrt{n(S + N_0)}$.

It follows that if we assume as indistinguishable the points that are in the spheres defined by the noise — whose radius is $\sqrt{nN_0}$ — the maximum number of signals distinguishable by the channel coincides with the maximum number of spheres of this radius that we can pack within the sphere of radius $\sqrt{n(S + N_0)}$ keeping them pairwise disjoint (see the figure).

This number coincides with the ratio of the volume of the large sphere to that of the small sphere. Now, it is known that the volume of the sphere of radius $r > 0$ in $\mathbb{R}^n$ is given by

$$V_n(r) = \frac{\pi^{\frac{n}{2}}}{\Gamma(\frac{n}{2} + 1)} r^n,$$

where $\Gamma(z) = \int\limits_0^\infty t^{z-1} e^{-t} dt$ is the gamma function. Thus, the quotient that interests us is the quotient of the radii of the spheres, raised to the nth power

$$m = \left(\frac{\sqrt{n(S + N_0)}}{\sqrt{nN_0}} \right)^n = \left(\frac{n(S + N_0)}{nN_0} \right)^{n/2} = \left(1 + \frac{S}{N_0} \right)^{n/2}.$$

As this is the total amount of different signals that we can send — being in principle all equally probable — the volume of information that can be sent when transmitting one of these signals through the channel during the time interval $[-T/2, T/2]$ is

$$\log_2 m = \frac{n}{2} \log_2 \left(1 + \frac{S}{N_0}\right) = \frac{N(W, T, \varepsilon, \epsilon)}{2} \log_2 \left(1 + \frac{S}{N_0}\right).$$

Consequently, for any fixed positive bandwidth W, the capacity of the channel results from dividing this value by T and taking $T \to \infty$,

$$C = \lim_{T \to \infty} \frac{N(W, T, \varepsilon, \epsilon)}{2T} \log_2 \left(1 + \frac{S}{N_0}\right) = W \log_2 \left(1 + \frac{S}{N_0}\right) \text{ [bits/s]},$$

which is known as the "Shannon–Hartley formula."

It follows that to increase the capacity of one of these channels, we simply need to increase the signal-to-noise ratio — for example, by increasing the signal power while limiting the noise power, which in mundane terms is equivalent to raising the tone of voice when we are in a noisy environment — or increase the bandwidth allowed by the channel.

Faster than Nyquist signaling

We have already mentioned H. Nyquist several times in this book. Specifically, we owe him the stability criterion for feedback systems. Regarding the transmission of signals, Nyquist observed that when a discrete signal $\{a_n\}_{n \in \mathbb{Z}}$ (in which each a_n represents an M-ary symbol and carries an energy E_s) is transmitted through a communication channel with bandwidth $W < \infty$ by linear modulation and using orthogonal pulses, the maximum transmission rate that the channel supports without inter-symbol interference (ISI) is precisely $2W$. Specifically, linear modulation consists of sending a signal of the type

$$s(t) = \sqrt{E_s} \sum_{n=-\infty}^{\infty} a_n h(t - nT),$$

where the function $h(t)$ represents a pulse and has bandwidth W. This signal is filtered and sampled by the receiver so that every T seconds one of the symbols a_n is retrieved. Therefore, the transmission speed of the symbols a_n carried by the signal is $\frac{\log_2 M}{T}$.

The orthogonality condition consists of assuming that

$$\int_{-\infty}^{\infty} h(t - kT)h(t - mT)dt = 0 \text{ for } n \neq m,$$

and this implies the absence of interferences between symbols during transmission. That is, it allows that a filter and a sampler in the receiver at instant kT retrieve the symbol a_k. Furthermore, Gaussian channel noise affects each retrieval independently.

Now, if we allow interferences between symbols, the bound $2W$ is not an insurmountable impediment. For a long time, it was assumed that this bound was absolute because no one knew how to handle the problem of interference between symbols. However, in 1975, the *Bell System Technical Journal* published an article by J. E. Mazo (1937–2013) in which it was proposed to accelerate the transmission speed by a factor of $1/\tau$ simply by sending the signal

$$s_\tau(t) = \sqrt{E_s} \sum_{n=-\infty}^{\infty} a_n h(t - n\tau T).$$

This signal has the same bandwidth as the signal $s(t)$ above, but the pulses are transmitted at speed $\frac{1}{\tau T} = \frac{1}{\tau}R > R = \frac{1}{T}$.

Next, he showed that for a certain pulse $h(t)$ and certain values $\tau < 1$, the error rate with optimal detection of the new method matches that of the classical Nyquist method.

At that time, eliminating the interference between symbols was not a computationally treatable problem, but this changed considerably in the 1980s. Today, there are very efficient implementations that solve the problem and are used, for example, in wireless mobile communications.

Wiener's and Shannon's Race for Information Theory

We know that Shannon developed his early work on communication theory under the influence of Wiener since he recognized it in the 1950s. For his part, Wiener was tremendously jealous of Shannon's mathematical work and, despite being much older than his competitor and already enjoying recognition as a great mathematician when Shannon published his communication theory, was very sensitive to the pressure caused by the work of others when they dealt with the same topic. By the mid-1940s, Wiener was aware that Shannon was making headway in his research on entropy and information. This made him feel pressured to close the publication of his own research on that subject. This finally happened when he published his book on cybernetics, in which he devoted an entire chapter to "Time series, information, and communication." As expected, his approach was eminently analog and included as a fundamental ingredient the use of ergodic theory, to which he had previously contributed. The problem of coding, which is central to Shannon's approach, was not addressed at all in *Cybernetics*.

It is known that Wiener had a conflict with Pitts in 1946 because, after leaving him a manuscript in which he explained his ideas about information, Pitts lost it with the rest of his luggage on a trip he made from New York to Boston, and it took five months to get it back. Wiener complained to the MIT authorities, stating that he had "lost priority on some important work" and that "One of my competitors, Shannon of the Bell Telephone Company, is coming out with a paper before mine."

All this shows that Wiener saw Shannon as a competitor and that he aspired to be recognized as the creator of information theory. When *Cybernetics* appeared in 1948 and garnered near-unanimous praise from readers and critics, Wiener calmed down. The aspiration for recognition, which was so evident in Wiener, did not seem that important to Shannon, and this had the effect that there never seemed to be a direct confrontation between the two. In the 1950s, they met on several occasions and published about information theory in the same scientific journals. At first, Shannon seemed to recognize Wiener's influence. But over time, Shannon moved away from that position, to the point that in an interview he had in the 1980s, he even affirmed his opinion that Wiener had never understood the substance of information theory:

> When I talked to Norbert, like in the 1950's and so on, I never got the feeling that he understood what I was talking about.

And, in another interview, he said:

> I don't think Wiener had much to do with information theory.
> He was not a big influence on my ideas there, though I once
> took a course from him.

In fact, there is considerable consensus that the true source of inspiration for Shannon was Hartley and Nyquist's work rather than Wiener's. At first, perhaps he thought that Wiener was part of what he was building, but then he saw that in reality they both had very different interests and that when they spoke, they did so about different things.

In the end, Shannon and Wiener created two different schools of thought. Shannon's led to the development of what we now call "information and coding theory" and Wiener, whose main promoter was Yuk Win Lee, created what we now call "statistical theory of communication." Obviously, both disciplines have many points in common, but they have different objectives and methods. Shannon's point of view focused on information encoding. In contrast, the methods developed by Wiener focused on the use of generalized harmonic analysis, ergodic theorems, and wave filter theory to deal with the problem of controlling the flow of information. Currently, both theories are considered important and form — so to speak — the heads and tails of the same coin.

Chronology

1894: Norbert Wiener is born in Columbia, Missouri (USA), son of Leo Wiener and Bertha Kahn.

1901: First trip to Europe.

1903: Wiener's first admission to the school. Before this, his education had been the responsibility of his father.

1906–1909: Tufts College. He graduated in philosophy, with a special mention in mathematics, at age 14.

1913: Doctor of Philosophy from Harvard University. Scholarship to visit Russell. He meets Hardy. He travels to Göttingen, where he meets Hilbert, Landau, etc.

1914: Start of the First World War. Wiener returns to the US.

1917: America enters the First World War. Wiener tries to enlist in the army but is not admitted due to his extreme myopia.

1919: Gets a job as a teacher in the mathematics department at MIT.

1920: Characterization of the body structure based on a single binary operation. Congress of Strasbourg. Definition of the Wiener–Banach spaces. He visits Frechet.

1920–1923: Mathematical foundation of Brownian motion. He publishes some papers.

1924: Solution of the Zaremba problem. Conflict with O. D. Kellogg.

1925: Conference in Göttingen on the uncertainty principle of signal theory. He is invited to stay the following academic year. He receives Born at MIT, with whom he writes an important article on quantum mechanics.

1926: Heaviside operational calculus. Causal operator concept. Distribution concept. Generalized solution of the telegraph equation. He marries Marguerite Engelmann.

1927: Barbara, Wiener's first daughter, is born.

1928: First works on Tauberian theorems.

1929: Full professor at MIT. Peggy, Wiener's second (and last) daughter, is born.

1930: Publishes "Generalized Harmonic Analysis" in *Acta Math.* Yuk Wing Lee's thesis.

1930–1931: He meets E. Hopf, whom he helps enter the mathematics department at MIT. Together they publish an important article on integral equations (in 1931). In it, the Wiener–Hopf equations are studied for the first time.

1931–1932: Visiting professor at the University of Cambridge. He lectures on the Fourier Integral and its applications at Trinity College and publishes his first monograph on this subject at Cambridge University Press. He participates in the ICM in Zurich as a representative of MIT.

1932: He publishes "Tauberian theorems" in *Ann. Math.* Professor at MIT Tauberian theorems.

1933: Böcher Prize. Elected member of the National Academy of Sciences. He meets Arturo Rosenblueth, who will be best friends with him for the rest of his life. He visits Paley, with whom he would work on the Fourier transform in the complex domain. Characterization of physically feasible filters. Paley's death in a skiing accident.

1934: Book with Paley.

1935: First patent with Yuk Wing Lee. (He would have two more in 1938.)

1935–1936: Travels to China, where he visits Y. W. Read at the Tsing Hua University in Beijing. He participates in the ICM in Oslo. He knows S. Mandelbrojt.

1937: Dohme Lecture at John Hopkins University on Tauberian theorems.

1938: Semi-centennial Conference of the AMS.

1940: Memorandum on a digital computer. He begins working with J. Bigelow on his project for the study of anti-aircraft batteries.

1941: He resigned as a member of the National Academy of Sciences.

1942: Drafting of the Yellow Peril: "Extrapolation, interpolation, and smoothing of stationary time series" (The report will be released in book form and for the general public in 1949.) Meets McCulloch.

1943: He meets Pitts.

1944: Creation, with J. von Neumann, of the "Teleological Society."

1946: Named doctor honoris causa by Tufts College. He attends the first three Macy Lectures. In December, he publishes his letter in The Atlantic Monthly.

1946–1950: He gets, together with Rosenblueth, a grant from the Rockefeller Foundation so that they can visit each other every year, one year at MIT and the next at the National Institute of Cardiology of Mexico.

1947: Congress in Nancy on harmonic analysis. Meets Freymann.

1948: First version of Cibernética. (The second version appears in 1961.)

1949: Lord & Taylor American Design Award.

1950: The human use of human beings. (The second version appears in 1954.)

1951: McCulloch and "his boys" move to the MIT. Electronics Research Laboratory. Wiener gives a Fulbright Lecture in Paris. He visits Spain, where he lectures in Madrid.

1952: Breaks with McCulloch. Alvarega Award from the Philadelphia College of Physicians. He teaches the Forbes-Haws Lectures at the University of Miami.

1953: Ex-prodigy. Summer school, with R. Fano and C. Shannon, on the mathematical problems of communication theory.

1955: Visiting professor in Calcutta.

1956: I am a mathematician. Conference in Japan. Summer School at UCLA (repeated in 1959, 1961, and 1963).

1957: Doctor *honoris causa* by the Grinnell College. Virchow Medal from the Rudolf Virchow Medical School.

1960: Lectures at the University of Naples (returns in 1962). He visits Russia. ASTME Research Medal. Emeritus Professor at MIT.

1964: National Medal of Sciences. God and Golem. He dies in Stockholm of a heart attack.

Doctoral Students of Wiener

Gleason Kenrick. "A New Method of Periodogram Analysis with Illustrative Applications," Thesis co-directed with Frank Hitchcock, 1927, MIT.

Carl Muckenhoupt. "Almost Periodic Functions and Vibrating Systems," Thesis co-directed with Philip Franklin, 1929, MIT.

Dorothy Weeks. "A Study of the Interference of Polarized Light by the Method of Coherency Matrices," 1930, MIT.

Yuk Wing Lee. "Synthesis of Electric Networks by Means of the Fourier Transforms of Laguerre's Functions," 1930, MIT.

Shikao Ikehara. "An Extension of Landau's Theorem in the Analytic Theory of Numbers," 1930, MIT.

Sebastian Littauer. "Applications of the Fourier Transform Theorem on the Exponential Scale," 1930, MIT.

James Estes. "The Lift and Moment of an Arbitrary Aerofoil-Joukovsky Potential," 1933, MIT.

Norman Levinson. "On the Non-Vanishing of a Function," 1935, MIT.

Henry Malin. "On Gap Theorems," 1935, MIT.

Bernard Friedman. "Analyticity of Equilibrium Figures of Rotation," 1936, MIT.

Brockway McMillan. "The Calculus of the Discrete Homogeneous Chaos," 1939, MIT.

Abe Gelbart. "On the Growth Properties of a Function of Two Complex Variables Given by its Power Series Expansion," 1940, MIT.

Colin Cherry. "On Human Communication: A Review, a Survey, and a Criticism," 1956, Imperial College.

Amar Gopal Bose. "A Theory of Nonlinear Systems," Thesis co-directed with Yuk Wing Lee, 1956, MIT.

Donald Brennan. "On the Pathological Character of Independent Random Variables," 1959, MIT.

William Stahlman. "The Astronomical Tables of Codex Vaticanusgraecus 1291," Thesis co-directed with Otto Neugebauer, 1960, Brown University.

Donald Tufts. "Design Problems in Pulse Transmission," Thesis co-directed with Yuk Wing Lee, 1960, MIT.

George Zames. "Nonlinear Operators for System Analysis," Thesis co-directed with Yuk Wing Lee, 1960, MIT.

Bibliography

Wiener's Publications: The following are works by Norbert Wiener that have been explicitly or implicitly mentioned in this biography. A complete list of his publications is available at [Ma] and on the website: https://archivesspace.mit.edu/repositories/2/resources/600.

[**1913**] On the rearrangement of the positive integers in a series of ordinal numbers greater than that of any given fundamental sequence of omegas, *Messenger of Math*, **3**.

[**1914**] A simplification of the logic of relations, *Proc. Cambridge Philos. Soc.* **27**, 387–390.

[**1915**] Studies in synthetic logic, *Proc. Cambridge Philos. Soc.* **18**, 24–28.

[**1920a**] A set of postulates for fields, *Trans. Amer. Math. Soc.* **21**, 237–246.

[**1920b**] The mean of a functional of arbitrary elements, *Ann. Math. (2)* **22**, 66–72.

[**1920c**] On the theory of sets of points in terms of continuous transformations, *G. R. Strasbourg Math. Congress*.

[**1921a**] The average of an analytical functional, *Proc. Nat. Acad. Sci. U.S.A.* **7**, 253–260.

[**1921b**] The average of an analytical functional and Brownian motion, *Proc. Nat. Acad. Sci. U.S.A.* **7**, 294–298.

[**1924**] Certain notions in potential theory, *J. Math. Phys.* **3**, 24–51.

[**1923a**] Note on a paper of M. Banach, *Fund. Math.* **4**, 136–143.

[**1923b**] Nets and the Dirichlet problem (with H. B. Phillips), *J. Math. Phys.* **2**, 105–124.

[**1923c**] Differential space, *J. Math Phys.* **2**, 131–174.

[**1923d**] Discontinuous boundary conditions and the Dirichlet problem, *Trans. Amer. Math. Soc.* **25**, 307–314.

[**1924a**] The quadratic variation of a function and its Fourier coefficients, *J. Math. Phys.* **3**, 72–94.

[**1924b**] The Dirichlet problem, *J. Math. Phys.* **3**, 127–147.

[**1924c**] Une condition nécessaire et suffisante de possibilité pour le problème de Dirichlet, *C. R. Acad. Sci. Paris* **178**, 1050–1053.

[**1924d**] The average value of a functional, *Proc. London Math. Soc.* **22**, 454–467.

[**1926a**] A new formulation of the laws of quantization for periodic and aperiodic phenomena (with M. Born), *J. Math. Phys.* **5**, 84–98.

[**1926b**] The operational calculus, *Math. Ann.* **95**, 557–584.

[**1929**] Appendix, in *Fourier Analysis and Asymptotic Series, Operational Circuit Analysis* (by V. Bush, Wiley, New York), 366–379.

[**1930**] Generalized harmonic analysis, *Acta Math.* **55**, 117–258.

[**1931**] Uber eine Klasse Singularer Integralgleichungen (with E. Hopf), S. B. Preuss. *Akad. Wiss*, 696–706.

[**1932**] Tauberian theorems, *Ann. Math.* **33**, 1–100.

[**1933**] *The Fourier Integral and Certain of Its Applications*, Cambridge Univ. Press, New York; reprint, Dover, New York, 1959.

[**1934**] Fourier transforms in the complex domain (with R. E. A. C. Paley), *Amer. Math. Soc. Colloq. Publ.*, Vol. 19, *Amer. Math. Soc, Providence, RI.*

[**1936a**] Sur les séries de Fourier lacunaires. Théorème direct (with Szolem Mandelbrojt), *C. R. Acad. Sci. Paris* **203**, 34–36.

[**1936b**] Séries de Fourier lacunaires. Théorème inverses, with Szolem Mandelbrojt, *C. R. Acad. Sci. Paris* **203**, 233–234.

[**1936c**] Gap theorems, *C. R. de Congr. Int'l. des Math.*

[**1938a**] The homogeneous chaos, *Amer. J. Math.* **60**, 897–936.

[**1938b**] The historical background of harmonic analysis, *Amer. Math. Soc.* Semicentennial Publications Vol. II, Semicentennial Addresses, *Amer. Math. Soc, Providence, R. I.*

[**1939a**] The ergodic theorem, *Duke Math. J.* **5**, 1–18.

[**1939b**] The use of statistical theory in the study of turbulence, *Nature* **144**, 728.

[**1943a**] Behavior, purpose, and teleology (with Arturo Rosenblueth and J. Bigelow), *Philos. Sci.* **10**, 18–24.

[**1943b**] The discrete chaos (with Aurel Wintner), *Amer. J. Math.* **65**, 279–298.

[**1945**] The role of models in science (with Arturo Rosenblueth), *Philos. Sci.* **12**, 316–322.

[1946a] The mathematical formulation of the problem of conduction of impulses in a network of connected excitable elements, specifically in cardiac muscle (with Arturo Rosenblueth), *Arch. Inst. Cardiol. Méxicana* **16**, 205–265.

[1946b] A scientist rebels, *Atlantic Monthly* **179**, 46.

[1948a] *Cybernetics, or Control and Communication in the Animal and the Machine*, Actualités Sci. Ind., no. 1053; Hermann et Cie., Paris; The M.I.T. Press, Cambridge, Mass. and Wiley, New York. Versión española: *Cibernética o El control y comunicación en animales y máquinas*, Metatemas **8**, Ed. TusQuets, 1985 (2ªEdición, 1998).

[1948b] An account of the spike potential of axons (with Arturo Rosenblueth, W. Pitts, and J. Garcia Ramos), *J. Comp. Physiol*, December.

[1949] *Extrapolation, Interpolation, and Smoothing of Stationary Time Series: With Engineering Applications*, The MIT Press, Cambridge, Mass.; Wiley, New York; Chapman & Hall, London.

[1950a] *The Human Use of Human Beings*, Houghton Mifflin, Boston, (paperback edition (Anchor) by Doubleday, 1954) (Edición en castellano: *Cibernética y sociedad*, Ed. Sudamericana, 1958).

[1950b] Purposeful and non-purposeful behavior (with Arturo Rosenblueth), *Philos. Sci*, October.

[1950c] Comprehensive view of prediction theory, Proceedings of the International Congress of Mathematicians, Cambridge, Mass., Vol. 2, pp. 308–321; *Amer. Math. Soc.*, Providence, R. L, 1952, Expository lecture.

[1950d] Entropy and information, *Proc. Sympos. Appl. Math.*, Vol. 2, *Amer. Math. Soc., Providence, R.I.*, 89.

[1951] Problems of sensory prosthesis, *Bull. Amer. Math. Soc.* **57**, 27–35.

[1953a] A new form for the statistical postulate of quantum mechanics (with Armand Siegel), *Phys. Rev.* **91**, 1551–1560.

[1953b] *Ex-prodigy: My Childhood and Youth, Simon and Schuster*, New York; The MIT Press, Cambridge, Mass. 1965 (also paperback edition by MIT Press), *Math. Rev.* **15** (1954), 277.

[1953c] The concept of homeostasis in medicine, *Trans. Stud. Coll. Physicians Phila. (4)* **20**, No. 3.

[1953d] The differential space theory of quantum mechanics (with Armand Siegel), *Phys. Rev.* **91**, 1551.

[1956] *I Am a Mathematician: The Later Life of a Prodigy*, Doubleday, Garden City, New York, (paperback edition by The MIT Press).

[1957] The prediction theory of multivariate stochastic processes. I. The regularity condition (with P. Masani), *Acta Math.* **98**, 111–150.

[1958a] The prediction theory of multivariate stochastic processes. II. The linear predictor (with P. Masani), *Acta Math.* **99**, 93–137.

[1958b] My connection with cybernetics. Its origin and its future, *Cybernetica*, 1–14.

[1958c] *Nonlinear Problems in Random Theory*, The MIT Press, Cambridge, Mass., and Wiley, New York.

[1959] *The Tempter*, Random House, New York.

[1960] Some moral and technical consequences of automation, *Science* **131**.

[1961] *Cybernetics*, Second edition, The MIT Press and Wiley, New York, (also paperback edition by MIT Press).

[1964] *God and Golem, Inc.: A Comment on Certain Points Where Cybernetics Impinges on Religion*, The MIT Press, Cambridge, Mass.

[1966] *Generalized Harmonic Analysis and Tauberian Theorems*, The MIT Press. (Reprint of the papers that appeared in *Acta Math.* in 1930 and *Ann. Math.* in 1932), Cambridge, Mass.

[1993] *Invention: The Care and Feeding of Ideas*, MIT Press.

Publications about Wiener, his life, and his work: In this list, we include a selection of articles and books whose central theme is to get closer to the work or life of Norbert Wiener.

[Ad] **D. R. Adams,** Potential and Capacity before and after Wiener, *Proc. of the Centenary Simposia on N. Wiener*, 63–80.

[AR1] **J. M. Almira, A. E. Romero,** "A note on Norbert Wiener contributions to Harmonic Analysis and Tauberian Theorems," in *Mathematical Models in Engineering, Biology and Medicine, American Institute of Physics Conference Proceedings*, **1124** (2009) 19–28. Proceedings of "*BVP2008*," American Institute of Physics, Melville, New York, 2009.

[Bo] **A. G. Bose,** Ten years with Norbert Wiener, Address presented at "A symposium on the Legacy of Norbert Wiener in Honor of the 100th Anniversary of his Birth," Cambridge, MA., 1994.

[Br] **Felix Browder, E. H. Spanier, M. Gerstenhaber (Eds.),** *Norbert Wiener: 1894–1964*. Providence, RI: Bulletin of the American Math. Society 72, Number 1, Part 2, 1966.

[CS] **F. Conway, J. Siegelman,** *Dark Hero of Information Age. In Search of Norbert Wiener, the Father of Cybernetics*, Basic Books, 2004, New York, NY.

[Fr] **H. Freudenthal,** Norbert Wiener, in *Dictionary of Scientific Biography*, Vol. XIV, Chas. Cribner's Sons, 1976, 344–347.

[Gr1] **I. Grattan-Guinness,** Wiener on the logics of Russell and Schröder, *Ann. Sci.* **32** (1975) 103–132.

[He1] **S. J. Heims,** *John Von Neumann and Norbert Wiener, From Mathematics to the Technologies of Life and Death.* MIT Press, Cambridge, Mass, 1980.

[JSS] **D. Jerison, I. M. Singer, D. W. Stroock, (Eds.),** The legacy of Norbert Wiener: A Centennial symposium, *Proc. Symposia in Pure Mathematics* 60, AMS, Providence, R.I., 1997.

[LLM] **Y. W. Lee, Norman Levinson, W. T. Martin (Eds.),** *Selected Papers of Norbert Wiener: Expository Papers by Y. W. Lee, Norman Levinson, and W. T. Martin,* The MIT Press, Cambridge, Mass., 1964.

[M] **V. Mandrekar,** Mathematical work of Norbert Wiener, *Notices of the AMS*, **42**(6) (1995), 664–669.

[Mar] **M. B. Marcus,** Book review: Dark hero of the information age: In search of Norbert Wiener the father of cybernetics, by F. Conway and J. Siegelman, Notices of the AMS, **53** (2006) 574–579.

[Ma] **P. R. Masani,** *Norbert Wiener (1894–1964),* Vita Mathematica, 5, Birkhäuser, Basel, Boston, Berlin, 1990.

[Ma1] **P. R. Masani (Ed.),** *Norbert Wiener: Collected Works.* Volumes I, II, III, IV. MIT Press, Cambridge, MA, and London, 1976, 1980, 1982 and 1986.

[Mo] **L. Montagnini,** *Harmonies of Disorder. Norbert Wiener: A Mathematician Philosopher of Our Time,* Springer, Cham, Switzerland, 2017.

Other publications related to Wiener: This list includes a selection of articles and books whose main subject is not Norbert Wiener but contains references to him or some aspect related to him or his work.

[AL] **J. M. Almira, U. Luther,** Inverse closedness on approximation algebras, *J. Math. Anal. Appl.*, **314**(1) (2006) 30–44.

[AR2] **J. M. Almira, A. E. Romero,** How distant is the ideal filter of being a causal one? *AEJM*, **3** (2008) 47–56.

[Ar] **M. A. Arbib,** *Brains, Machines, and Mathematics,* Springer-Verlag, New York, NY, 1987.

[Ha] **O. Hallonsten,** *Big Science Transformed,* Palgrave Macmillan, Switzerland, 2016.

[He2] **S. J. Heims,** *The Cybernetics Group,* The MIT Press, Cambridge, Mass, London, England, 1991.

[Ka] **D. Kaiser (Ed.),** *Becoming M.I.T.: Moments of Decision,* The MIT Press, Cambridge, Mass, 2012.

[Le] **Y. W. Lee,** *Statistical Theory of Communication,* John Wiley and Sons, New York, NY, 1960.

[Lu] **J. Lützen,** Heaviside's operational calculus and the attempts to rigorize it, *Arch. Hist. Exact. Sci.* **21** (1979) 161–200.

[Mi] **D. A. Mindell,** *Between Human and Machine. Feedback, Control and Computing before Cybernetics,* The Johns Hopkins University Press, Baltimore and London, 2002.

[Pe] **B. J. P. Peters,** Betrothal and betrayal: The Soviet translation of Norbert Wiener's early cybernetics, *Int. J. Commun.* **2** (2008), 66–80.

[Th] **C. W. Therrien,** The Lee-Wiener legacy, *IEEE Signal Processing Magazine* **19**(6) (2002) 33–44.

[Re] **C. Reid,** *Courant,* Copernicus, New York, NY, 1996.

[U1] **S. Ulam,** *Adventures of a Mathematician,* University of California Press, Berkeley, CA, 1991.

[Wl] **K. L. Wildes, N. A. Lindgren,** *A Century of Electrical Engineering and Computer Science at M.I.T., 1882–1982,* The MIT Press, Cambridge, Mass, 1985.

Other publications cited in this book: We include several articles and books implicitly or explicitly mentioned in this biography, many of whose authors are people directly related to Norbert Wiener or his work.

[Ba] **C. Barus,** "Military Influence on the Electrical Engineering Curriculum Since World War II," in *IEEE Technology and Society Magazine,* **6**(2) (1987) 3–9.

[B] **M. Born,** The statistical interpretation of quantum mechanics, *Nobel Lecture,* December 11, 1954.

[D] **J. H. Davis,** Stability of linear feedback systems, Ph.D. Thesis, MIT, 1970.

[LMMP] **J. Y. Lettvin, H. R. Maturana, W. S. McCulloch, W. H. Pitts,** "What the frog's eye tells the frog's brain," in *The Mind: Biological Approaches to its Functions,* William C. Corning, Martin Balaban (Eds.), 1968, 233–258.

[MC] **C. M. Madrid Casado,** De la equivalencia entre la mecánica matricial y la mecánica ondulatoria, *La Gaceta de la RSME* **10**(1) (2007) 103–128.

[MP] **W. S. McCulloch, W. H. Pitts,** A logical calculus of the ideas immanent in nervous activity, *Bull. Math. Biophy.* **5** (1943) 115–133.

[Mi1] **J. Mikusinski,** *Operational Calculus,* Pergamon Press, London, England, (1959).

[Mi2] **J. G. M.**, Hypernumbers, Studia Mathematika LXXVII (1983) 3–16 (Reprint of the original paper, that was published by his own means by Mikusinski in 1944).

[Ne1] **J. von Neumann,** *Mathematical Foundations of Quantum Mechanics,* Princeton University Press, Princeton, New Jersey, 1955.

[O] **S. Okamoto**, A simplified derivation of Mikusinski's operational calculus. *Proc. Japan Acad.,* **55A**(1) (1979) 1–5.

[Pu] **M. I. Pupin,** *From Immigrant to Inventor*, Cosimo, Inc., New York, NY, 2005.

[Ra] **W. R. Ransom**, *One Hundred Mathematical Curiosities.* Portland, ME: J. Weston Walch, 1955.

[Ro] **A. Rosenblueth,** *Mind and Brain; a Philosophy of Science*, MIT Press, Cambridge, Mass, 1970.

[S] **J. M. Sánchez Ron,** *El poder de la ciencia*, Ed. Crítica, Barcelona, Spain, 2011.

[Sc] **L. Schwartz,** Théorie des distributions, Hermann, Paris, Vol. I, 1950; Vol. II, 1951.

[Sha1] **C. E. Shannon,** A mathematical theory of communication, *Bell Syst. Tech. J.,* **27**(3) (1948) 379–423.

[Sha2] **C. E. Shannon,** "Communication in the Presence of Noise," in *Proc. IRE* **37**(1) (1949) 10–21.

[ShaWe] **C. E. Shannon, W. Weaver,** *The Mathematical Theory of Communication*, University of Illinois Press, Urbana, Illinois, 1949.

[Sl] **D. Slepian,** "On bandwidth," in *Proc. IEEE,* **64**(3) (1976) 292–300.

[WR] **A. N. Whitehead, B. Russell** (1910, 1912, 1913) *Principia Mathematica*, 3 vols, Second edition, Cambridge University Press, Cambridge, 1925 (Vol. 1), 1927 (Vols. 2, 3).

[WC] **S. Winograd, J. D. Cowan,** *Reliable Computation in the Presence of Noise*, The MIT Press, Cambridge, Mass, 1963.

[YO] **K. Yosida, M. J. A., and S. Okamoto,** A note on Mikusinski operational calculus, *Proc. Japan Acad.* **56**, Ser. A (1980) 1–3.

Other publications: Here, we include both introductory texts on history (of mathematics or science in general), physics, engineering, etc., as well as research or dissemination articles whose contents extend or clarify any aspect dealt with in this book.

[Ab] **N. Abramson,** *Information Theory and Coding*, McGraw Hill, New York, San Francisco, Toronto, London, 1963.

[An] **J. B. Anderson,** *Bandwidth Efficient Coding*, IEEE Series on Digital & Mobile Communication, Wiley-IEEE Press, Hoboken, New Jersey, 2017.

[AnJ] **J. B. Anderson, R. Johnnesson,** *Understanding Information Transmission,* IEEE Press Understanding Science & Technology Series, Wiley-IEEE Press, Hoboken, New Jersey, 2006.

[AF] **M. Alonso, E. J. Finn,** *Fundamental University Physics: Quantum and Statistical Physics Volume III,* Addison Wesley, Reading, Mass, 1968.

[AS] **J. M. Almira, J. C. Sabina de Lis,** *Hilbert. Matemático fundamental,* en La matemática en sus personajes, 31, Ed. Nivola, Madrid, Spain, 2007.

[Al] **J. M. Almira,** Cuerdas vibrantes y calor: la génesis del análisis de Fourier. Matematicalia, **4**(1) (2008).

[As] **W. Aspray,** *John Von Neumann and the Origins of Modern Computing,* The MIT Press, Cambridge, Mass, 1990.

[At] **P. Atkins,** *Four Laws That Drive the Universe,* Oxford University Press, Oxford and New York, 2007.

[Bo] **N. Bourbaki,** *Elements of the History of Mathematics,* Springer, Berlin and Heidelberg, 1994.

[Br] **V. M. Brodianski,** *Móvil perpétuo antes y ahora,* Mir, Moscú, URSS, 1990.

[Cu] **G. P. Curbera,** *Mathematicians of the World Unite! The International Congress of Mathematicians — A Human Endeavor,* AK Peters, Wellesley, Mass, 2009.

[F1] **J. A. Fleming,** *Fifty Years of Electricity. The Memories of an Electrical Engineer,* Wireless Press, London and New York, 1921.

[Fe] **A. Fernández-Rañada,** *Heisenberg. De la incertidumbre cuántica a la bomba atómica nazi,* en Científicos para la Historia (Serie Mayor) 1, Ed. Nivola, Madrid, Spain, (2008).

[Ga] **G. Gamow,** *Thirty Years that Shook Physics. The Story of Quantum Theory,* Doubleday & Co. Inc., New York, 1966.

[GNV] **C. Gerthsen, H. O. Kneser, H. Vogel,** *Física,* Springer-Verlag & Ed. Dossat, Madrid, Spain, 1979.

[GPS] **H. J. Glaeske, A. P. Prudnikov, K. A. Skòrnik,** *Operational Calculus and Related Topics,* Chapman & Hall/CRC, Boca Raton, Florida, 2006.

[Gr2] **I. Grattan-Guinness (Ed.),** *From the Calculus to Set Theory 1630–1910: An Introductory History,* Princeton University Press, Princeton and Oxford, 2001.

[GR] **M. de Guzmán, B. Rubio,** *Integración: Teoría y técnicas,* Alhambra, Madrid, Spain, 1979.

[Kh] **A. I. Khinchin,** *Mathematical Foundations of Information Theory,* Dover, New York, NY, 1957.

[Kr] **S. Krantz,** *A Panorama of Harmonic Analysis*, The Carus Mathematical Monographs 27, Mathematical Association of America, Providence, Rhode Island, 1999.

[Kr] **H. Kragh**, *Quantum Generations: A History of Physics in the Twentieth Century*, Princeton University Press, Princeton, New Jersey, 2002.

[Leo] **A. León-García**, *Probability and Random Processes for Electrical Engineering*, Addisson-Wesley, Upper Saddle River, New Jersey, 1994.

[Les1] **S. W. Leslie,** *The Cold War and American Science. The Military-Industrial-Academic Complex at M.I.T. and Stanford.* Columbia University Press, New York, 1993.

[Les2] **S. W. Leslie**, Profit and loss: The military and MIT in the postwar era. *Historical Studies in the Physical and Biological Sciences*, **21** (1) (1990) 59–85.

[Na] **P. J. Nahin**, *Oliver Heaviside. The Life, Work and Times of an Electrical Genius of the Victorian Age*, The Johns Hopkins University Press, Baltimore and London, 2002.

[Ne2] **J. von Neumann,** *The Computer and the Brain.* Yale University Press, New Haven and London, 1958.

[O] **A. V. Oppenheim, A.S. Willsky,** *Signals and Systems*, Prentice Hall, Upper Saddle River, New Jersey, 1982.

[Pi] **J. P. Pier,** *Histoire de l'intégration. Vingt-cinq siècles de mathémathique*, Masson, Paris, France, 1996.

[Pr] **I. Prigogine**, *The Birth of Time and Eternity*, Shambhala Pubns, Boulder, Colorado, 2000.

[Pre] **D. Preston**, *Before the Fallout: From Marie Curie to Hiroshima*, Walker & Company, New York, NY, 2005.

[Q] **M. Quesada Pérez**, El radar y la Segunda Guerra Mundial, *Revista Española de Física*, **18**(2) 2004, 58–61.

[SS] **E. B. Saff, A. D. Snider,** *Fundamentals of Complex Analysis*, Prentice-Hall, Englewood Cliffs, New Jersey, 1976.

[ScSa] **E. D. Schneider, D. Sagan,** *Into the Cool: Energy Flow, Thermodynamics, and Life*, The University of Chicago Press, Chicago and London, 2006.

[So] **D. J. de Solla Price,** *Little Science, Big Science.* Columbia University Press, New York, 1963.

Name Index

Subject Index

Printed in the United States
by Baker & Taylor Publisher Services